1812

ALSO BY WALTER R. BORNEMAN

A Climbing Guide to Colorado's Fourteeners
(with Lyndon J. Lampert)

Alaska: Saga of a Bold Land

1812

THE WAR THAT FORGED A NATION

WALTER R. BORNEMAN

HARPER
PERENNIAL

HARPER ⬤ PERENNIAL

Grateful acknowledgment is made to Warden Music for permission to quote from "The Battle of New Orleans," words by Jimmy Driftwood, copyright © Warden Music Co., Inc. Used by permission. All rights reserved.

A hardcover edition of this book was published in 2004 by HarperCollins Publishers.

P.S.™ is a trademark of HarperCollins Publishers.

HarperCollins books may be purchased for educational, business, or sales promotional use. For information please write: Special Markets Department, HarperCollins Publishers, 10 East 53rd Street, New York, NY 10022.

FIRST HARPER PERENNIAL EDITION PUBLISHED 2005.

Maps by Eric Janota

Designed by Nancy B. Field

The Library of Congress has catalogued the hardcover edition as follows:

Borneman, Walter R.
  1812 : the war that forged a nation / Walter R. Borneman.
  —1st ed.
    p. cm.
  Includes bibliographical references (p.) and index.
  ISBN 0-06-053112-6
  1. United States—History—War of 1812. 2. United States—History—War of 1812—Influence. I. Title.

  E354.B66 2004
  973.5'2—dc22   2004047261

ISBN-10: 0-06-053113-4 (pbk.)
ISBN-13: 978-0-06-053113-3 (pbk.)

    12 13 14 ❖/RRD 20 19 18 17

For my mother

Barbara Lucille Parker Borneman
(1927–1956)

CONTENTS

Illustrations follow page 176.

MAPS

ACKNOWLEDGMENTS

At first glance, a history of the War of 1812 may seem to be far afield from my previous writings on western history. In truth, as a youngster growing up in Ohio, I cut my teeth on tales of the American Revolution, the later battles along the Great Lakes, and the settlement of the Northwest Territory. When childhood fascination turned to academic study, I quickly came to realize that whatever happened west of the Mississippi had its roots in the eastern rivers and forests that had first captured my imagination.

My chief goals here are to present a readable history of the War of 1812, place it in the context of America's development as a nation, and emphasize its importance as a foundation of America's subsequent westward expansion. Though frequently overlooked between the Revolution and the Civil War, the War of 1812 did indeed span half a continent—from Mackinac Island to New Orleans and Lake Champlain to Horseshoe Bend—and it set the stage for the conquest of the continent's other half.

For those who wonder how I could write a history of the War of 1812 based in Colorado, I must thank the Denver Public Library, the Penrose Library of the University of Denver, and the Arthur Lakes Library of the Colorado School of Mines. I am also grateful to the research assistance of Fadra Whyte at the University of Pittsburgh and Christopher Fleitas at Yale University, and the cartographic skills of Eric Janota at National Geographic Maps.

In addition to newspapers and the *Annals of Congress*, many primary sources from this period are increasingly available in published form. These include the personal papers of such key figures as John Quincy Adams, Burr, Clay, Gallatin, Hamilton, Harrison, Jackson, Jefferson, Madison, Monroe, and Wilkinson. John C. Fredriksen's *War of 1812 Eyewitness Accounts* is the key to finding primary sources from lesser-known participants on both sides. Recent publication of *The Naval War of 1812: A Documentary History* has also placed a host of primary sources at one's fingertips. There have been many scholarly histories of the war or its phases over the years, but one of the most recent and clearly the best is Donald R. Hickey's *The War of 1812: A Forgotten Conflict*. Professor Hickey's insight and voluminous footnotes are a treasure trove and form bedrock for any serious study of the period. Other essential secondary sources include J. Mackay Hitsman's *The Incredible War of 1812* from the British and Canadian perspective; Robert Remini's biographies of Jackson and Clay; and Arsene Latour's *Historical Memoir of the War in West Florida and Louisiana* and Robert McAfee's *History of the Late War in the Western Country*—both published shortly after the conflict. Theodore Roosevelt's *The Naval War of 1812* and Alfred Thayer Mahan's *Sea Power in its Relations to the War of 1812* remain steady stalwarts.

As always, my favorite part of the research was being in the field. My beloved grandparents, Walter and Hazel Borneman, may have started it all by taking me to Brock's Monument at Queenston Heights at the age of four. In ensuing years, I walked the decks of "Old Ironsides," looked out across Put-in Bay, visited Presque Isle, and crossed the straits of Mackinac. More recently, my wife, Marlene, and I toured Fort McHenry, Chesapeake Bay, and Lake Champlain. Jim Gehres and Jean McGuey joined us on a whirlwind return to Queenston Heights, Lundy's Lane, and Chippawa, and Jim and Gail Fleitas extended us cordial southern hospitality for a reenactment of the Battle of New Orleans. This book makes number two with two of the best, my editor, Hugh Van Dusen, and my agent, Alexander Hoyt. I'm very glad to be a part of the team!

INTRODUCTION

THE WAR THAT
FORGED A NATION

In some respects, it was a silly little war—fought between creaking sailing ships and inexperienced armies often led by bumbling generals. It featured a tit-for-tat, "You burned our capital, so we'll burn yours," and a legendary battle unknowingly fought after the signing of a peace treaty. In the retrospect of two centuries of American history, however, the War of 1812 stands out as the coming of age of a nation.

In June 1812 a still infant nation of eighteen loosely joined states had the audacity to declare war on the British Empire. Such indignation was fired by resentment over years of British high-handedness on the high seas and envious yearnings by Americans west of the Appalachians for more territory. But a good part of the country, mostly the New England states, thought there was far more to lose by pulling the lion's tail than a handful of ships and impressed sailors. New Englanders did not take kindly to or go out of their way to support what they called "Mr. Madison's war."

Indeed, Great Britain might have easily crushed the upstarts if it had brought its full weight to bear on the matter, but the British were preoccupied with Napoleon's maneuvers in Europe, and the American war in North America remained a sideshow for its first two years. During that time the American navy proved its mettle and found an icon as the USS *Constitution*, "Old Ironsides," sent two first-rate British frigates to the bottom. On land, despite Henry Clay's boast that the militia of Kentucky alone was capable of conquering Canada, two years of invasion attempts on three fronts failed. When the tide appeared to turn in America's favor, it was because a twenty-seven-year-old lieutenant named Oliver Hazard Perry ran a flag boasting "Don't Give Up the Ship" up his flagship's mast and with a motley collection of hastily built ships chased the British navy from Lake Erie.

Within two months of Perry's victory, however, Napoleon met his first waterloo at Leipzig, and victorious Great Britain was free to swing a battle-hardened hand at its American cousins. Suddenly the young American nation was no longer fighting for free trade, sailors' rights, and as much of Canada as it could grab, but for its very existence as a nation. In 1814 Great Britain threatened assaults against all corners of the United States: from Canada via Niagara Falls and Lake Champlain, into the heartland of the mid-Atlantic states from Chesapeake Bay, and against that door to half a continent, New Orleans. For a time it looked as if the British Empire might regain its former colonies.

With Washington in flames, only a valiant defense at Fort McHenry saved Baltimore from a similar fate. The greatest American victory in 1814, however, may have come not on the battlefield or the high seas, but at the negotiating table. Somehow, with British armies arrayed along its borders and a British blockade locking up its ports, the United States managed to sign a peace treaty on Christmas Eve, 1814, that preserved its preexisting boundaries, even if it made no reference to one of the war's most egregious causes.

The news of the Treaty of Ghent reached Washington only after Andrew Jackson had assembled a tattered force of army regulars, backwoods militia, and bayou pirates and bested the pride of British regulars. Strategically, the Battle of New Orleans was of no military significance to the war, but politically it came to fill a huge void in the American psyche—not only propelling Andrew Jackson to the presidency, but also affirming a strong, new sense of national identity.

During the War of 1812, the United States would cast aside its cloak of colonial adolescence and—with more than a few bumbles along the way—stumble forth onto the world stage. After the War of 1812, there was no longer any doubt that the United States of America was a national force to be reckoned with. But in the beginning, all of this was very much in doubt.

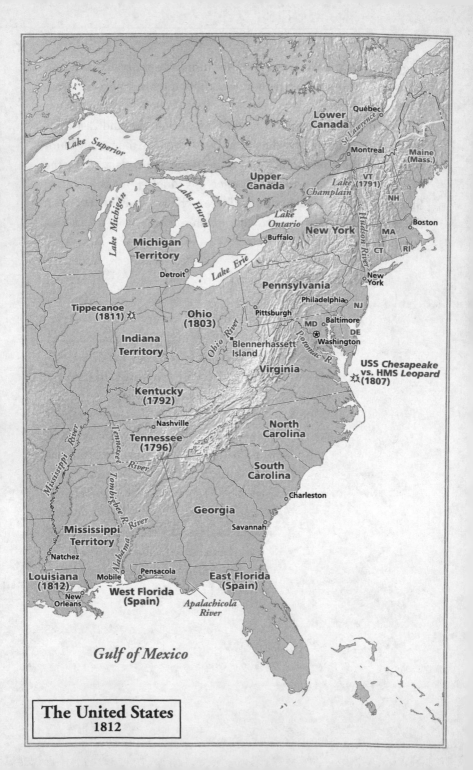

Lower Canada

Québec

St. Lawrence

Montreal

Maine (Mass.)

Lake Superior

Upper Canada

Lake Champlain

VT (1791)

NH

Lake Michigan

Lake Huron

Lake Ontario

Buffalo

New York

Hudson River

Boston

MA

CT

RI

Michigan Territory

Lake Erie

New York

Detroit

Pennsylvania

Tippecanoe (1811)

Ohio (1803)

Pittsburgh

Philadelphia

NJ

Indiana Territory

Ohio River

Blennerhassett Island

MD

Baltimore

Washington

DE

Potomac R.

USS *Chesapeake* vs. HMS *Leopard* (1807)

Virginia

Kentucky (1792)

Mississippi River

Tennessee River

Nashville

Tennessee (1796)

Tennessee River

North Carolina

South Carolina

Tombigbee R.

Charleston

Georgia

Savannah

Mississippi Territory

Alabama River

Natchez

Louisiana (1812)

Mobile

Pensacola

New Orleans

West Florida (Spain)

East Florida (Spain)

Apalachicola River

Gulf of Mexico

The United States
1812

BOOK ONE
Drumbeats
(1807–1812)

Many nations have gone to war in pure gayety of heart, but perhaps the United States were first to force themselves into a war they dreaded, in the hope that the war itself might create the spirit they lacked.

—*Henry Adams,* History of the United States During the Administrations of Jefferson and Madison, *1889*

To Steal an Empire

In the early twilight, the swollen waters of the Ohio River swept a wooden flatboat up to a landing on a small, tree-covered island. On the river's east bank lay the western reaches of the state of Virginia; on the west, the shores of the state of Ohio, now, in the spring of 1805, barely two years old. The flatboat was much grander than the normal river craft that floated by or landed here. Indeed, its owner had commissioned its recent construction in Pittsburgh, and he himself described it as a "floating house, sixty feet by fourteen, containing dining room, kitchen with fireplace, and two bedrooms, roofed from stem to stern. . . ."[1]

The flatboat belonged to Aaron Burr. With jet-black eyes, a silken tongue, and the refined dress to match the accoutrements of his vessel, Burr cast a larger shadow than his diminutive height suggested. For four years, he had been the proverbial heartbeat away from the presidency, but once he had also been just one particular heartbeat away. Why the recent vice president of the United States came to make this journey down the Ohio River evidences just how tenuous the American union still was in 1805, and that the very last thing it should have come to contemplate was another war with Great Britain.

In the presidential election of 1800, there were as yet no strictly organized political tickets. Prior to the Twelfth Amendment, the

Constitution merely ordained that the person receiving the highest number of electoral votes be declared president and the second highest, vice president. Party electors were supposed to withhold a vote or two from the agreed-upon vice presidential candidate, thus assuring the election of their presidential favorite.

Such informality didn't work very well. In fact, so many Federalist electors withheld votes from John Adams's running mate in 1796 that Republican Thomas Jefferson ended up with the second highest number of votes and the vice presidency. (Jefferson's Republicans were the liberal predecessors of the Jefferson–Jackson Democrats and not the "Grand Old Party" of Abraham Lincoln.) To avoid such a result in 1800, Republican vice presidential candidate Aaron Burr obtained Jefferson's assurance that no southern elector would drop a vote for Burr, but that Burr would arrange for a Republican elector from Rhode Island—supposedly a solid Jefferson state—to withhold one vote for Burr. That strategy backfired when the Federalists proceeded to win Rhode Island, and the remaining Republican electors cast the identical number of votes for president and vice president.

Thus in only the nation's fourth presidential election, Thomas Jefferson handily defeated incumbent John Adams, but imagine Jefferson's surprise when his vice presidential running mate received the same number of electoral votes as he, and the election was declared a tie. With Jefferson and Burr each receiving seventy-three votes, the election went to the House of Representatives, where the contest was suddenly not between Federalist and Republican, but between Republican and Republican.

Vice presidential candidate Burr professed allegiance to Jefferson, but made no outright disclaimer of the higher office. Indeed, there were plenty of whispers in Burr's ear to suggest that the higher office was his for the taking. New England Federalists, who were rarely as unified in anything as they were in their opposition to Thomas Jefferson, actively courted Burr, vastly preferring the New York lawyer—Republican though he might be—to the Virginia planter.[2]

Not all Federalists felt that way, of course. Alexander Hamilton for one was appalled at the possibility of Burr becoming president. Four years before he would die by Burr's ducling pistol, Hamilton wrote: "There is no doubt but that upon every virtuous and prudent calculation Jefferson is to be preferred. He is by far not so dangerous a man and he has pretensions to character."[3] Among other things, Hamilton probably feared that Burr might come to take over the Federalist Party that Hamilton clearly viewed as his own exclusive route to the presidency.

In the House of Representatives, the Federalists controlled six states, the Republicans eight. Two states were undecided. A simple majority of nine was needed to elect either Jefferson or Burr president. For a turbulent six weeks, the electoral balloting and the intra-party intrigue continued. Certain Federalists and Republicans friendly to Burr clung to the hope that they might be able to swing three states into the Federalist column and make Burr president. Finally, after some backroom concessions obtained from Jefferson through Alexander Hamilton, James A. Bayard of Delaware—the undecided state's lone vote in the House of Representatives—voted for Jefferson to give him the required nine states. Aaron Burr would spend four years being a heartbeat away from the presidency, but he lost it by the single heartbeat of James Bayard.

Both Jefferson and Burr were quick to say that each bore no hard feelings toward the other, but more than a few Republicans noted how far Burr had been tempted to stray to the Federalists, and, likewise, the Federalists knew how close they had come to getting him. The result was that both sides came to view Burr as something of a leaf willing to be blown by whatever political winds offered the promise of greater glory. For Jefferson's part, he would soon prove that he hadn't meant that line about "no hard feelings" after all.[4]

So Aaron Burr became vice president of the United States in March 1801. By most accounts he served a distinguished term, taking seriously his charge to preside over the United States Senate and tarnishing his reputation only through his duel with

Alexander Hamilton. Even by the standards of 1804, it is difficult to grasp that a sitting vice president of the United States should fight a duel, let alone kill his opponent, but in truth Thomas Jefferson had been determined to rid himself of Burr long before the public uproar over the duel.

On February 25, 1804, before the days of political conventions, congressional Republicans met to nominate Jefferson for a second term. New York Governor George Clinton received the lion's share of vice presidential votes ahead of Kentuckian John Breckinridge and a handful of favorite sons. The incumbent vice president, Aaron Burr, received none. Four years before, Burr had been a single heartbeat away from the presidency; now he could not muster the support of a single heart that thought him worthy of the *vice* presidency. It was a humiliating repudiation.[5]

Burr's first thought was to redeem himself by winning the governorship of New York. When he failed miserably in that attempt, there was little to do but finish out his term as vice president and head west. His direction was chosen with a great deal of forethought.

At the time, the ink was scarcely dry upon the Louisiana Purchase treaty. Louisiana was a vast chunk of territory—roughly the western drainage of the Mississippi River—that France had ceded to Spain in 1763 as part of its North American concessions following the Seven Years' War. Four decades later, Napoleon managed to bully Spain into returning it, but Spanish administrators in Louisiana retaliated by revoking the "right of deposit" enjoyed by the fledgling United States. Negotiated by the Washington administration in 1795, the right of deposit permitted Americans to export farm products and other goods through New Orleans. When the right of deposit was revoked, it was tantamount to choking the commerce of all of the United States west of the Appalachians.

By 1803 Jefferson wanted Louisiana in part to resolve the right of deposit issue. Napoleon desperately needed American cash to fund his European conquests. Even if Napoleon sold

Louisiana, a steady stream of clandestine reports seemed to suggest that he might be able to have his cake and eat it, too. The very acquisition of Louisiana by the United States might, wishful thinking suggested, prompt the split of the still tenuous American union between the New England and mid-Atlantic states more interested in European commerce, and the western states more in tune with Napoleon's swaggering militarism. Then Louisiana and *all* of the Mississippi Valley might be persuaded to become a French ally. With fifteen million American dollars in his vaults, Napoleon might still be able to retain effective control of Louisiana. Now there was a touch of Napoleonic optimism of the sort that would find him at the gates of Moscow a decade hence![6]

In this early period of American history, the idea of states seceding from the Union was not nearly as cataclysmic as it would be deemed two generations later. Indeed, people in all parts of the country maintained an intense, first-line loyalty to their respective home states. As early as 1792, Secretary of State Thomas Jefferson warned President George Washington that southern states might secede in opposition to Alexander Hamilton's national banking system. And even as the Senate debated the ratification of the Louisiana Purchase, Senator John Breckinridge of Kentucky threatened that if the treaty was rejected, "the western states would immediately secede and form a separate country."[7]

While some hailed Jefferson's purchase of Louisiana a diplomatic masterpiece, others argued that the United States should simply have seized the territory over the right of deposit issue. There were others who were convinced that no matter the present transaction, should Napoleon be victorious in Europe, the United States would be called upon to defend the new acquisition from his rapacious territorial appetite. But when Jefferson dispatched 450 troops under the command of Brigadier General James Wilkinson to occupy New Orleans in the wake of the purchase, it was not out of fear of the French, but rather that the Spanish who still controlled the city might resist. Nonetheless, Wilkinson and newly appointed Governor William Claiborne took possession of the Crescent City without incident on December 20, 1803.[8]

James Wilkinson was either the most despicable scoundrel in American history or the victim of the worst press ever. Most accounts support the former. Wilkinson had known Aaron Burr for more than a quarter of a century, ever since their paths crossed as young officers serving under Benedict Arnold during his ill-fated 1775 attack on Quebec during the opening round of the American Revolution. Two years later as a deputy adjutant general, Wilkinson helped to draft the terms of surrender for Burgoyne's British army at Saratoga—a fact that Wilkinson was always quick to relate.

Despite his long tenure in the United States Army, however, Wilkinson took an oath of allegiance to Spain as early as 1787. To the Spanish, he was known as "Agent 13." Such duplicity aside, by 1792 Wilkinson was a brigadier general, and the following year commanded a wing of General "Mad" Anthony Wayne's army at the Battle of Fallen Timbers. When Wayne died in 1796, James Wilkinson was left as the ranking general in the U.S. Army, a position he would occupy for the next fifteen years.

But Wilkinson was seldom happy. In May 1802 Wilkinson wrote Aaron Burr of his disillusionment and bitterness over lack of congressional support for the military in general and machinations with his rank and salary in particular. Small wonder Wilkinson was disposed to line his pockets with Spanish gold. Burr may not have known of that, but he certainly knew Wilkinson to be disgruntled by the actions of eastern politicians while envisioning himself as the guardian of the young nation's frontier.[9]

Another to whom Wilkinson complained was Alexander Hamilton. Wilkinson assured Hamilton that he had explored "with military eyes . . . every critical pass, every direct route, & every devious way between the Mexican Gulph [sic] & the Tennessee River." Wilkinson intimated that he was tired of President Jefferson's pacifism and that to wear his sword "without active service is becoming disreputable."[10] Despite such criticisms, Wilkinson somehow managed to remain Jefferson's favorite solider. Perhaps it had something to do with the fact that he wrote Jefferson lengthy letters filled with descriptions of flora and fauna.

Over the objections of Secretary of War James McHenry, who would soon have a fort near Baltimore named after him, Hamilton had previously recommended Wilkinson's promotion to major general. In doing so, Hamilton urged McHenry that "we ought certainly to look to the possession of the Floridas and Louisiana—and we ought to squint at South America."[11]

Wilkinson had scarcely settled into New Orleans when he had a series of late-night conversations with Vincente Folch, the governor of Spanish West Florida. Agent 13 gave the Spaniard helpful suggestions about how to keep the Americans out of the Floridas and lessen American influence in newly acquired Louisiana. Shortly afterward, Wilkinson—with his most recent Spanish retainer payment jingling in his pockets—set out on a trip back east.[12]

On May 23, 1804, in the wake of his repudiation as vice president and would-be governor of New York, Aaron Burr received a late night caller, who made a point that no one should know of his visit. Never mind the fact that Burr and his secretive guest were old friends. With Burr's political power apparently destroyed, what could he and James Wilkinson possibly have to talk about?

No record remains of their conversation. Subsequent events, however, suggest that the man who had just told the Spanish how to keep American influence at bay in the Louisiana Territory was now on Aaron Burr's doorstep suggesting to the disgruntled politico a way to steal an empire away from both American and Spanish influences. Louisiana also meant Mexico and the entire Southwest, and Wilkinson had with him maps of Texas and the Spanish Southwest.[13]

Because he was a man determined to be on the winning side at all costs of honor, Wilkinson had also written to Alexander Hamilton a few weeks before his clandestine meeting with Burr. The "destinies of Spain" in the entire Southwest, Wilkinson assured Hamilton, were in "the hands of the U.S."[14] Several months later, Hamilton was dead, the consequence of a bullet fired by Aaron Burr in the culmination of a quarter of a century of political rivalry. Hamilton's friends demanded a criminal indictment. Burr kept a low profile, but he was not without a plan.

On August 6, 1804, the British ambassador to the United States, Anthony Merry, sent a report of an astonishing communication to the British foreign secretary. "I have just received an offer from Mr. Burr the actual vice president of the United States (which Situation he is about to resign)," wrote Merry, "to lend his assistance to His Majesty's Government in any Manner in which they may think fit to employ him, particularly in endeavouring to effect a Separation of the Western Part of the United States from that which lies between the Atlantick [sic] and the Mountains, in it's [sic] whole Extent."[15]

Meanwhile, the rightful owner of that territory appointed General James Wilkinson—who had suggested to anyone who would listen or grease his palm with gold that all of Louisiana was ripe for the picking—governor of Upper Louisiana. Significantly, as he moved upriver to St. Louis and assumed his gubernatorial duties, Wilkinson continued to hold his military rank. Despite criticisms in Congress that civil and military authority should not repose in the same person, the general/governor would be waiting for Aaron Burr to pay a return visit.

Such was the background to the journey that found Aaron Burr touching shore at the landing on Blennerhassett Island in the spring of 1805. If Burr was a particularly distinguished visitor to the island, his host and hostess were themselves rather extraordinary. Harman Blennerhassett traced his roots to English nobility. His ancestors immigrated to Ireland during the fourteenth century, and Harman was born there on October 8, 1764, or perhaps 1765, the youngest of three sons and six daughters. "My father and mother," Harman later wrote, "were never agreed as to which year I was born."[16] As the third son, he chose law as his profession, but the death of his father and two older brothers soon made him heir to a considerable fortune.

In 1796, while visiting Captain Robert Agnew, the lieutenant governor of the Isle of Man, Blennerhassett was dispatched aboard to escort Agnew's daughter home from school. He was thirty-one, tall, learned, but somewhat awkward. Margaret Agnew was a captivatingly beautiful eighteen. She, too, was tall. With blue eyes and auburn hair, she was a skilled horsewoman and spoke and wrote both French and Italian. Enough said. Blennerhassett cast aside all notion of his fiduciary role and persuaded her to marry him on the spot. There was only one problem: she was also his niece. Upon their return to the Isle of Man, the reception was less than cordial, and Blennerhassett promptly sold his estate in Ireland for a reported $160,000—then a considerable sum—and sailed for New York with his elegant bride, arriving there on August 1, 1796.[17]

The newlyweds spent some time first in New York and then in Philadelphia, but soon headed west across the Alleghenies. They found their way via keelboat down the Ohio to Marietta, then the gateway to much of the Northwest Territory. From the heights above Marietta, Blennerhassett discovered an island in the Ohio River that at once boasted both the lushness of Ireland and a regal setting from which to command an empire. In March 1797 he purchased two hundred acres on the island and was soon ensconced there with Margaret, who quickly bore him three children. By the time of Burr's arrival, the centerpiece of Blennerhassett Island was a fifty-four by thirty-eight-foot mansion with two thirteen-foot-wide and thirty-seven-foot-long porticos that connected to symmetrical wings. Perhaps Burr mused that not even Thomas Jefferson's executive mansion in Washington City had such fine appendages as yet.[18]

On this spring evening, Harman and Margaret Blennerhassett greeted their guest warmly. While Burr was an anathema to many easterners who still mourned the loss of Alexander Hamilton, here in the West men viewed such matters as dueling quite differently. Burr was treated as the celebrity he certainly was. Once again, there is no complete record of the conversations that took place that evening, but subsequent events suggest that Burr appealed to Blennerhassett's ego as well as his pocketbook.

Burr soon continued down the Ohio bound for New Orleans, but he called again at Blennerhassett Island on his return upriver. Afterward Blennerhassett wrote Burr looking for money in exchange for a sale or lease of his island. "In either way, if I could sell or lease the place, I would move forward with a firmer confidence in any undertaking which your sagacity might open to profit and fame," suggested Blennerhassett. "Having thus advised you of my desire and motives to pursue a change of life, to engage in any thing which may suit my circumstances, I hope, sir, you will not regard it indelicate in me to observe to you how highly I should be honored in being associated with you in any contemplated enterprise you would permit me to participate in."[19]

That desired association almost left Blennerhassett high indeed—swinging high from the end of a rope. Blennerhassett would later testify that he had no clear understanding of what venture Burr had in mind, but he nonetheless agreed to assist Burr in the construction of certain boats and to permit a company of men to gather on Blennerhassett Island.

Another stop Burr made in May 1805 and again in August upon his return from New Orleans was with Andrew and Rachel Jackson in Nashville. The Jacksons welcomed Burr as a man of the West. Jackson himself had fought his share of duels, and he could well relate with talk of driving the Spanish from not only the Floridas but also Texas. Jackson viewed such talk as just good American boosterism—the sort of thing that would come to be called "manifest destiny." If Jackson saw Burr's dark side and suspected his true motives, he gave no inkling of it until November 10, 1806, an uncomfortable week or so after Jackson, too, had accepted money from Burr for the construction of two boats.

On that date Captain John A. Fort arrived in Nashville from New Orleans and stayed with the Jacksons. Over the course of the evening—apparently thinking Jackson more of the insider than he was—Fort went on at some length about Burr's plans to seize

New Orleans and make it the gateway to a great southwestern empire. Suddenly disturbed, Jackson sent a series of letters, including one to President Jefferson, suggesting that something was amiss along the Mississippi. Jefferson at first was confused by Jackson's missive. Was the belligerent frontiersman merely suggesting a war with Spain, or was there in fact something treasonable behind Burr's actions?[20]

Initially Burr was arrested and arraigned before a grand jury in Frankfort, Kentucky, on charges of raising troops for illegal purposes. Burr's lawyer was a young up-and-comer named Henry Clay. Finding that "no violent disturbance of the Public Tranquility or breach of the laws" had occurred, the grand jury dismissed the charges. Nonetheless, Jackson wrote Burr of his suspicions, and Burr hastened to Nashville to reassure Jackson. Whatever he said, it must have mollified Jackson, because on December 22, 1806, Burr left Nashville with the two boats he had purchased from Jackson and floated down the Cumberland, planning to rendezvous with the boats and men coming from Blennerhassett Island.[21]

Scarcely had Burr disappeared downstream, however, than a presidential proclamation reached Nashville announcing that President Jefferson believed a vast military conspiracy was under way in the West and calling for the arrest of all those involved, principally his former vice president. Jefferson was at least partly correct, of course, but the man who had urged him to take such action and the man to whom Jefferson now turned to save the Union was none other than James Wilkinson.

Much has been written about Wilkinson's true motives. His detractors far outnumber his supporters. Why he chose to double-cross Burr, when Wilkinson himself had first sung the siren's song of empire in Burr's willing ear, is debatable, but in no small measure it was motivated by Wilkinson's desire always to be on the winning side. Perhaps he saw the futility of a handful of men in flatboats stealing an empire. Undoubtedly he bemoaned the potential loss of his regular Spanish retainer. Going to great pains to cover his own conspiratorial tracks, Wilkinson wrote a subordinate: "By letters found here, I perceive the plot thickens; yet all but those concerned sleep profoundly. My God! What a situation has our country reached."[22]

The upshot was that Aaron Burr was arrested again and taken to Richmond, Virginia, to stand trial for treason. He was acquitted only because Chief Justice John Marshall chose to preside over the trial himself and narrowly define treason under the Constitution as requiring an overt act of war. Marshall had political debts to Burr, and it was not his most shining moment as a jurist. General Wilkinson was a star witness for the government and played the self-righteous defender of America's frontier. Agent 13's reputation emerged from the trial as muddled as ever.[23]

While Burr was on trial, Harman Blennerhassett spent fifty-three days in a Richmond jail awaiting his own fate and ruing the day that he and Margaret had welcomed such a wily guest to Blennerhassett Island. With Burr's acquittal, all charges were dropped against Blennerhassett, but his beloved mansion overlooking the Ohio River had been overrun and damaged by local militia in the wake of his arrest.[24]

Meanwhile, Andrew Jackson, who had himself come uncomfortably close to Burr's web, wrote a friend: "I am more convinced than ever that treason never was intended by Burr, but if ever it was you know my wishes that he may be hung."[25]

So Louisiana and the Mississippi Valley still belonged to the United States. The fledgling American union remained intact, but the plotting of a disgruntled politician and an unscrupulous general had shown how tenuous it was. Within a short time, growing pains along its entire border from British Canada to Spanish Florida, as well as on the high seas, would push the country into war. When that war came, Aaron Burr had faded into the shadows, but many of the others who had played a role in the machinations of what came to be called the "Burr Conspiracy" would be center stage, including Andrew Jackson, Henry Clay, and the nefarious General Wilkinson.

FIRST BLOOD AT SEA

Two days before a grand jury indicted Aaron Burr for treason, the thirty-eight-gun frigate USS *Chesapeake* sailed eastward from the mouth of Chesapeake Bay, bound for station in the Mediterranean. When she was barely beyond the three-mile territorial limit and in international waters, the fifty-gun British frigate HMS *Leopard* pulled alongside her. The *Leopard*'s captain demanded the right to search the *Chesapeake* and detain questionable British deserters. In the wake of the broadside that followed, the United States was almost plunged into what might have come to be called the War of 1807.

Such conflict on the high seas was certainly not new to the fledgling United States. Since winning its independence, the country had been forced repeatedly to tangle not only with the era's superpowers of Great Britain and France, but also with the likes of tribute-demanding pirates. In addition to general trade concerns, a nagging issue was the impressment of American citizens into the British navy. How degrading to the young nation to have its ships stopped, its citizens seized and then made to serve in virtual imprisonment on board His Majesty's warships.

Oppressive as impressment was to most, the British government viewed the policy as essential to its mastery of the seas. British law permitted any able-bodied male subject to be drafted into immediate service in the Royal Navy—anytime, anywhere.

This meant that a British captain who was shorthanded could put into any friendly or neutral port throughout the world and send a "press gang" ashore to round up likely recruits. Frequently this was done amid the pubs and brothels, but press gangs also called on commercial vessels flying the Union Jack or the flags of neutral countries, including the United States.

Legally, such impressed recruits had to be British subjects, but practically, British citizenship throughout a worldwide dominion had many shades of gray. When, for example, did a British immigrant to the United States cease to be a British subject? The United States had naturalization laws, of course, but the British crown held to simpler criterion: "Once an Englishman, always an Englishman."

While Great Britain's practice of impressment was longstanding, its use escalated as British manpower needs increased during the Napoleonic wars. Between 1803 and 1812 at least five thousand and perhaps as many as nine thousand sailors were impressed from the decks of American ships and forced to serve in the Royal Navy. Perhaps as many as three-quarters were bona fide American citizens.

In 1807 Secretary of State James Madison decried the entire practice of impressment as "anomalous in principle, grievous in practice, and abominable in abuse." His demands that the practice cease met only with contempt from British authorities, particularly the foreign secretary, who scoffed: "The Pretension advanced by Mr. Madison that the American Flag should protect every Individual sailing under it . . . is too extravagant to require any serious Refutation."[1] So impressment incidents continued.

In the spring of 1807, a British naval squadron made itself at home in the waters off the Virginia Capes at the mouth of Chesapeake Bay. The ships dropped anchor to await the emergence of two French warships that had been driven into the bay by storms. The United States was ostensibly neutral in the conflict between Great Britain and Napoleon, and such hospitality to visiting warships by neutral countries was a recognized principle of international law. Many British captains, however, made a habit of singing "Rule Britannia" just a little too stridently no matter where they made port.

While HMS *Melampus* was anchored in Hampton Roads, several of its crew—the exact number is in doubt—chose to desert, making off with the captain's gig in the process. Three of the deserters promptly enlisted for service aboard the American frigate *Chesapeake*, then being outfitted at Norfolk for a patrol in the Mediterranean Sea against the Barbary pirates. When the British consul at Norfolk demanded their return, the request wound up the chain of command until Secretary of State James Madison himself denied the request, eventually saying that he believed these particular seamen to be American citizens.

Affidavits obtained later showed that William Ware and John Strahan (Strachan in some reports) had been born in Maryland and that Daniel Martin had been brought to Massachusetts as an indentured servant at the age of six. All three claimed that they had been impressed against their will into the Royal Navy. The British claimed that the three had enlisted. Even if the latter was true, Madison took the legal position that the enlistment of American citizens by a belligerent power when their own country was neutral was void. The three seamen in question were to remain on the *Chesapeake*.[2]

While they did so, there were rumors about Norfolk that others aboard the *Chesapeake* might be British deserters. British captains reported their frustrations with the situation up their own chain of command to Vice Admiral George Cranfield Berkeley, commanding His Majesty's ships throughout North America. Berkeley dispatched orders from his headquarters in Halifax, Nova Scotia, that any ship that encountered the *Chesapeake* at sea should stop her, show his order, and conduct a search for deserters.

Berkeley's order was carried to the British squadron off the Virginia Capes by HMS *Leopard*, a fifty-gun frigate under the command of Captain Salusbury P. Humphreys. The *Leopard* arrived off the Virginia Capes on June 21, 1807. The next day, the *Chesapeake* sailed from Norfolk.

The *Chesapeake* was under the nominal command of Captain Charles Gordon, but was serving as the flagship of Commodore James Barron, who had already distinguished himself in previous operations in the Mediterranean. While it was Gordon's responsibility to have a trained and able crew and to take such steps as might be necessary to defend his vessel, he was all too happy to defer to Barron, particularly in light of the events that soon transpired. Certainly Gordon was aware of rumors that the captain of the *Melampus* had threatened to remove the three seamen from the *Chesapeake* by force. Nonetheless, he anticipated no trouble and did nothing to prepare for it as his ship weighed anchor.[3]

About nine o'clock on the morning of June 22, the *Chesapeake* sailed past the *Melampus* at anchor in Lynnhaven Bay, without incident. By noon her lookouts sighted another ship off Cape Henry. This proved to be the *Leopard*. The British frigate began to shadow the *Chesapeake*, but neither Gordon nor Barron grew uneasy, even after both ships passed into international waters later that afternoon. Then, about ten miles off Cape Henry, the *Leopard* closed with the *Chesapeake* and hailed her. Hearing that Captain Humphreys desired to send a messenger aboard, Barron ordered the *Chesapeake* to heave to. When the messenger stepped onto the deck before Barron, it was to deliver Admiral Berkeley's ultimatum that the ship be searched for British deserters. It was hardly the action one routinely took on board a neutral warship in international waters, but Berkeley's orders were evidently ringing loudly in Humphreys's ears.[4]

Incredulous at such British audacity, Barron firmly refused the search request and ordered the messenger returned to the *Leopard*. As the messenger departed, Barron ordered Gordon to bring the *Chesapeake* to general quarters "without beat of drum" so as not to "provoke hostility if none was designed."[5] He need not have worried. Gordon's crew was poorly trained, and the order began to be executed only at a disorganized and belabored pace.

There followed some minutes of shouted conversation between Barron and Humphreys through speaking trumpets—which included Barron's repeated assertions that he could not understand what was being said. Finally the *Leopard* fired a warning shot and then a full broadside into the unprepared *Chesapeake*. All the while, the *Chesapeake*'s guns remained covered and, in the words of Theodore Roosevelt in his *The Naval War of 1812*, "so lumbered up that she could not return a shot."[6]

Unable to return fire, Barron struck his colors. Three of his crew lay dead; eight others were seriously wounded. Ten more, including Barron himself, were slightly wounded. Humphreys dispatched a boarding party to conduct the search and examine the 370-some members of the *Chesapeake*'s crew. The British removed the disputed Ware, Strahan, and Martin, as well as a John Wilson, alias Jenkin Ratford, who was both a deserter and a British subject. Before sailing, Ratford had boasted of his escape in the bars of Norfolk, and the British later took great delight in hanging him in Halifax.

After the four men were on board the *Leopard*, Humphreys declined to take the *Chesapeake* as a prize, sensing correctly that such action would be an even more flagrant act of war than that which he had already committed with his broadside. He ordered the *Leopard* to join the British squadron in Lynnhaven Bay. Meanwhile, as dusk fell, the *Chesapeake* got up sail and limped back to Hampton Roads with three and a half feet of water in her hold.[7]

The sight of dead and wounded American sailors being taken off the *Chesapeake* caused an instant uproar around Hampton Roads. Indignation spread as Barron's report of the incident reached the secretary of the navy in Washington. Soon, demands for immediate redress resounded in all corners of the country. Talk of war was rampant. The sectional rivalries over the federal banking system and Jefferson's purchase of Louisiana melted into the background. As he carefully considered an American response, Thomas Jefferson correctly observed that the British "have often enough, God knows, given us cause of war before; but it has been on points which would not have united the nation. But now they have touched a chord which vibrates in every heart."[8]

In fact, the *Leopard*'s broadside into the *Chesapeake* did more to unify the country in 1807 than the rhetoric of a hundred war hawks would in 1812. Even Federalist John Adams joined in the thunder. "No nation," asserted Adams, "can be Independent which suffers her Citizens to be stolen from her at the discretion of the Naval or military officers of another."[9] Had Congress been in session or Jefferson been inclined to summon it, a declaration of war might have ensued in the summer of 1807.

As it was, most in the country—including Jefferson—were too consumed that summer by Aaron Burr's trial in Richmond to make good boasts of war. In the end, Jefferson chose to ban all British ships from American ports and resort to an economic embargo. Meanwhile, James Barron was made the scapegoat. He was tried before a naval court of inquiry for neglect of duty, convicted, and relieved of any command for five years. It was a lesson the American navy would remember. Never again would an American captain be so slow to unlimber his guns. As for the *Chesapeake*, she would live to fight another day.

It helped to calm the American clamor for war in the wake of the *Chesapeake* affair that His Majesty's Government quickly conceded that Captain Humphreys's actions had been grossly in violation of international law. Not even the British were willing to force enough of a stiff upper lip to claim the right to search a sovereign man-of-war in international waters. Admiral Berkeley himself was recalled. Prompt restitution, however, was another matter. So, too, was any change in British impressment policies.[10]

What eventually occurred was a tit-for-tat of the sort that would be repeated later when the issue was one of the burning of capitals and not just impressed sailors. On May 1, 1811, almost four years after the *Chesapeake* affair and while American and British ministers were still trying to agree to some measure of restitution, a press gang from the frigate HMS *Guerrière* impressed an American sailor from a coastal vessel off New York harbor. Captain John Rodgers and the American frigate USS *President* sailed in search of the *Guerrière* but in the dark of night mistook the smaller sloop HMS *Little Belt* for the frigate.

On May 16, 1811, the *President* opened fire. While the *Little Belt* gamely exchanged a couple of broadsides with her, the frigate escaped "scot-free while the sloop was nearly knocked to pieces."[11] Nine British sailors were killed and twenty-three wounded. It was as flagrant an act of war as had been the *Leopard*'s firing on the *Chesapeake*, and while it momentarily quickened the drumbeats of war, it did provide a trade-off with which finally to settle the *Chesapeake* affair.

WAR HAWKS AND
TIPPECANOE

There was one distinct region of the country where talk of war with Great Britain did not cool in the aftermath of the *Chesapeake* affair. This was the western frontier. In the years following the Revolutionary War, the country's adventuresome had spilled westward across the Appalachian Mountains into the headwaters of the Ohio and Tennessee rivers. By and large, they were the free spirits of the age—fiercely independent, politically liberal, and driven by the challenge of taming a wilderness. These were the men and women who had been quick to welcome Aaron Burr—not because he whispered of treason against the United States, but because he dreamed of greater empires extending northward to Canada and southward to Mexico.

In the wake of the Revolution, the federal government exerted its sovereignty and extinguished the claims of individual states to lands west of the Appalachians that had previously been held by Great Britain. In 1787—even before the Constitution was ratified—Congress looked at the vast triangle between the Ohio River, the Great Lakes, and the Mississippi, and created the Northwest Territory. The Northwest Ordinance of 1787 decreed that out of this sweep of land could be carved not fewer than three and not more than five new states. When five thousand men of voting age had settled in one particular section of the territory, that region could become a separate territory, elect a legislature, and send a nonvoting delegate to Congress. When that territory's population reached sixty thousand, it could become a state.

Two events built on the legal framework of the Northwest Ordinance and encouraged the rapid settlement of the Northwest Territory. In 1794 pro-British Federalists negotiated Jay's Treaty with Great Britain. While principally concerned with normalizing relations and promoting trade between the two countries, the treaty also required the British to abandon all their military posts on the American side of the Great Lakes. This served to make good Great Britain's Revolutionary War concessions and better delineate the boundary between the Northwest Territory and Canada.

Also in 1794, Revolutionary War veteran General "Mad" Anthony Wayne led an army of regulars and militia north from the site of Cincinnati along the future Indiana-Ohio border and defeated a combined Indian force that included Delaware, Miami, Ottawa, Potawatomi, and Shawnee at the Battle of Fallen Timbers. (Because all contemporary and most secondary accounts of the War of 1812 use the term "Indians," it has been retained in collective references rather than the currently accepted "Native Americans.") Among the Shawnee present at the battle was an up-and-coming young chief named Tecumseh. Wayne followed up his victory by negotiating the Treaty of Greenville in 1795, whereby the assembled tribes ceded major chunks of territory. The result was that most of Ohio and eastern Indiana was opened to settlement. Harman and Margaret Blennerhassett were among those who joined the rush to these lands.

By 1803, Ohio was the seventeenth state. Indiana hoped that it would be next. Despite Wayne's victory at Fallen Timbers, however, this rapid expansion caused repeated conflicts between Indians and incoming white settlers. While some confrontations were strictly local, both the United States and Great Britain long suspected each other of surreptitiously provoking Indian unrest against the other as part of its broader frontier policy. It was hardly that clear-cut. But as the drumbeats of war with Great

Britain grew louder, *Niles' Weekly Register* summed up the feelings along the American frontier. "We have had but one opinion as to the cause of the depredations of the Indians," the paper asserted, "which was, and is, that they are instigated and supported by the British in Canada, any official declaration to the contrary notwithstanding."[1]

And most westerners made no secret of the fact that they coveted Canada itself. American interest in Canada was nothing new. Aaron Burr and General Wilkinson had both been young lieutenants at Benedict Arnold's side during the 1775 Revolutionary War attempt to seize Quebec. Many postwar leaders—Federalist and Republican alike—assumed that Canada was a plum that would someday drop into the American union. (Even after the Civil War, there were still those in the American Congress who argued for the annexation of Canada as compensation for Great Britain's less than neutral assistance to Confederate raiders such as CSS *Alabama* and CSS *Shenandoah*.)

These twin issues of Indian unrest and a lust for additional territory beyond the Great Lakes heated the pot of war sentiment on the western frontier. Thoughts of quelling Indian influence for good and ousting Great Britain from Canada became the rallying cry for Henry Clay and his close-knit circle of political compatriots who came to be called the "war hawks."

A Virginian by birth, Henry Clay had already brushed against history as Aaron Burr's lawyer after his first arraignment in Frankfurt, and as the Kentucky representative of Andrew Jackson's trading enterprises. After Kentucky became a state, Clay served two piecemeal terms in the U.S. Senate before being elected to the U.S. House of Representatives in 1810. He was immediately elected Speaker of the House.

Since he was not yet thirty-five and a freshman representative, Clay's election to the speakership was evidence both of his ability to articulate the position of the war hawks and the fact that the election of 1810 produced 59 freshmen members of Congress out of 142— an unusually high turnover rate even then. It helped, too, that the

Republicans—who were beginning to call themselves Democratic-Republicans—controlled the House of Representatives 108 to 36 over the Federalists, and the Senate by an even greater margin of 30 to 6.

Clay's fellow war hawks included Richard Mentor Johnson of Kentucky, Felix Grundy of Tennessee, John C. Calhoun of South Carolina, George M. Troup of Georgia, and Peter B. Porter of New York, the last a state that by no small coincidence bordered on Canada. Nationalistic in policy, prompt with a dueling pistol when polite discussion failed, the war hawks were the young Turks of the era: too young to remember the devastation of the last war, and certain of their invincibility in the next.[2]

Clay's talents quickly made the heretofore largely ceremonial speakership a position of true power. He unceremoniously ordered crusty John Randolph of Virginia to remove his dog from the House floor—something no previous Speaker had dared to do—but he also exerted power to direct debate, interpret rules, and, most important, to pack key committees with his fellow war hawks. When one report of the Committee on Foreign Relations came to the House floor, Randolph—perhaps still chafing over the expulsion of his dog—declared in exasperation: "we have heard but one word—like the whip-poor-will, but one eternal monotonous tone—Canada! Canada! Canada!"[3] Under Clay's leadership, the Twelfth Congress became known to history as the War Congress, and Clay ensured that it stayed on the road to war.

One westerner and Clay ally was already well on the road to war. His name was William Henry Harrison, and he was governor of Indiana Territory. He had been born on February 9, 1773, in Charles City County, Virginia. His father, Benjamin, had signed the Declaration of Independence and twice been elected governor of Virginia. Young William received his army commission from President George Washington and was soon aide-de-camp to General "Mad" Anthony Wayne. In that capacity he laid out the plan of march that culminated in Wayne's victory at the Battle of

Fallen Timbers. Wayne's advance north from Cincinnati had been plodding and cautious, but with strength in numbers that made certain that the defeats that had befallen previous American military forays into the Northwest Territory would not be repeated.

After commanding Fort Washington at the site of present-day Cincinnati, Harrison resigned his commission in 1798 and accepted President John Adams's appointment as secretary of the Northwest Territory. When the territory of Indiana was carved out of the larger territory in 1800, Harrison was appointed its governor.[4]

Over the next decade, Governor Harrison engineered a series of land deals with various Indian tribes by frequently promising aid against their traditional enemies in exchange for land, but then promising competing tribes rights to the same lands. It was a classic strategy of divide and conquer, and the map of the Northwest Territory was soon covered with the cross claims of various Indian nations and the checkerboard townships marking the steady advance of white settlers. If such tactics were duplicitous, they were common on the American frontier, and in Harrison's mind, the end clearly justified the means. The end he sought to secure was ownership of the rich prairie land along the Wabash—what he termed "one of the fairest portions of the globe"—as "the seat of civilization, of science, and of true religion."[5]

Harrison's shrewdest bargain was the Treaty of Fort Wayne. In September 1809 Harrison summoned tribal representatives to meet him at the fort, some two hundred miles northeast of the territorial capital of Vincennes. After the ink marks dried, Delaware, Miami, Kickapoo, Potawatomi, and others had given up three million acres between the Wabash and White rivers, essentially the western portion of central Indiana. The transaction cost the United States government about a third of a penny per acre in annuities and goods. The land office would soon be selling the same lands to farmers for two dollars per acre. "I think," wrote Harrison to Secretary of War William Eustis, "upon the whole that the bargain is a better one than any made by me for lands south of the Wabash."[6]

There was, however, one Indian tribe conspicuously absent from the Treaty of Fort Wayne. The Shawnee and their principal chief, Tecumseh, had not been invited to the gathering. Harrison reported to the Indiana legislature that this was because "it has never been suggested that they could plead even the title of occupancy to the lands which were then conveyed to the United States."[7] Harrison thought that the Miami had the much stronger claim to these lands, and that the Shawnee were but recent, nomadic arrivals.

The Shawnee were certainly no strangers to migration. Originally they had lived in the western Carolinas and Georgia. Wars with the Creeks forced them northward in two groups to the Wabash Valley of Indiana and the Wyoming Valley of Pennsylvania. The Pennsylvania group was soon forced westward into Ohio, and in 1768 Tecumseh was born there on the Mad River near present-day Dayton. He had considerable contact with white settlers—even courting a settler's daughter for a time according to some sources.[8] The exact number of Tecumseh's siblings and the order of his birth in his family are in doubt, but what is clear is that he had a younger brother called Tenskwatawa, who quickly became known as the Prophet. Together the two brothers formed a dynamic team: Tecumseh as the military leader and visionary statesman promoting a powerful Indian confederacy, and the Prophet as the spiritual leader energizing the movement with a charismatic mysticism.

Tecumseh's dream of an Indian confederacy to halt the onslaught of white settlers was certainly not new. An Ottawa chieftain named Pontiac had tried it unsuccessfully two generations before in the wake of the French and Indian War (1763), and the dream would have its last gasp in the fires of Wounded Knee (1890), three generations hence. But at the time, Tecumseh's plan looked promising. The Americans were, after all, less than united. The *Chesapeake* affair aside, most of New England cared little about what went on in the West. Great Britain still controlled Canada on the north, and Spain continued to dole out funds to

General Wilkinson, a.k.a. Agent 13, in the hope of hanging on to the Floridas and Texas and somehow expelling the spike of an American Louisiana that had been thrust between them. Perhaps Tecumseh could unite the Indian nations of the Mississippi and Ohio River valleys before they were pushed westward onto the plains.

While so many of his brethren were meeting with Harrison at Fort Wayne, Tecumseh was in fact on a mission in the South among the Shawnee's old adversaries, the Creek, to rally support for his confederation. Upon his return to the banks of the Wabash, he denounced the land concessions and vowed to keep the ceded lands from being surveyed and settled. When Harrison heard of this, he invited Tecumseh and up to thirty warriors to meet with him in Vincennes and present evidence of any claims they might have to the ceded lands.

Tecumseh arrived in Vincennes in August 1810 not with thirty warriors but with four hundred. Greatly outnumbered, Harrison swallowed hard and listened as Tecumseh furiously denounced the whites of the "seventeen fires"—the seventeen states—in the walnut grove outside Harrison's Grouseland estate. Despite his superiority in numbers, Tecumseh was focused on the larger victory of the confederacy and knew that one battle here before it was in place would serve only to unite the white frontier against him. He withdrew after exchanging only rhetoric.

In the summer of 1811 Tecumseh appeared again at Vincennes for another meeting, this time bringing three hundred warriors with him. Not to be caught unawares again, Harrison assembled six hundred militia to meet him. Once more the conference produced only rhetoric, but Tecumseh may well have boasted too openly about his plans for a grand confederation. He asked Harrison to delay any surveys or settlement of the disputed lands until his return in the spring. Then, while the majority of his followers returned to the encampment of Prophetstown on the lands in question, Tecumseh and about twenty warriors floated off down the Wabash to once again visit the Creek in the South.[9]

Harrison had had ample opportunity to assess his adversary, and he was more than impressed. "If it were not for the vicinity of the United States," Harrison reported to Secretary of War Eustis as Tecumseh floated south, "he [Tecumseh] would perhaps be the founder of an empire that would rival in glory that of Mexico or Peru. No difficulties deter him. His activity and industry supply the want of letters. For four years he has been in constant motion. You see him today on the Wabash and in a short time you hear of him on the shores of Lake Erie or Michigan, or on the banks of the Mississippi and wherever he goes he makes an impression favorable to his purposes. He is now upon the last round to put a finishing stroke to his work."[10]

Before that finishing stroke could be written, Harrison was determined to strike a blow of his own in Tecumseh's absence and at the very least establish a presence in the disputed lands. His determination was bolstered by the deployment of the Fourth U.S. Infantry from Pittsburgh to the governor's disposal at Vincennes. So as Tecumseh went south to rally greater support for his confederacy, Harrison moved north. On September 26, 1811, his force of more than 900 officers and men, including 250 Fourth Infantry regulars, 60 Kentucky cavalrymen, and 600 Indiana militia, marched up the Wabash prepared to repel any attack.

Harrison had not forgotten the lessons he had learned as a young lieutenant at the side of General "Mad" Anthony Wayne en route to Fallen Timbers. A company of riflemen took the point 150 yards out in front and was followed by a mounted troop. Then came the main infantry column with companies of flankers deployed to the right and left sides. Another mounted troop brought up the rear. Always one to drill his troops relentlessly, Harrison had taught the flankers to turn outward in a line and fall backward slowly to the protection of the main column in the event of an attack. Such tactics were rather foreign to the roughshod Indiana militia, but they provided strength and security akin to a Roman phalanx as Harrison's army moved north.

Taking the time to establish a strong defensive camp each

evening, the column took two weeks to travel sixty-five miles to the high ground where Terre Haute now stands. There they built Fort Harrison, so named, Harrison later reported, at the insistence of his officers. Ahead lay the territory disputed by Tecumseh and the site of Prophetstown near the confluence of the Wabash and Tippecanoe rivers. Established about 1805, Prophetstown had quickly become a symbol of Indian resistance and the rallying point of both Tecumseh's confederacy and the Prophet's incantations.

At some point, Harrison made up his mind to destroy Prophetstown. Perhaps he reckoned to do so even before departing Vincennes. With Tecumseh away to the south and his militia bolstered by Colonel John P. Boyd's trained regulars, there was no time like the present to rid the Indiana frontier of this center of Indian resistance. Still, the governor could not appear to be too much of the aggressor. Fortunately for Harrison, with Tecumseh away the Prophet played right into his hands.

By November 5, 1811, Harrison's army had moved within a few miles of Prophetstown. A delegation of Indians met with Harrison and requested a parley. They invited Harrison to camp his force about one mile west of Prophetstown on a triangular-shaped, oak-covered knoll that dropped off into prairie wetlands about three miles below the mouth of the Tippecanoe River. Almost thirty years later, Harrison's political critics would still be finding fault with his willingness to camp at the Indians' suggested site; yet he found it a defensible position and deployed his men to make the most of it. Wagons and the main tents were grouped in the middle of the triangle and companies spread out around them in order of battle. Harrison sent his troops to sleep that evening with guns loaded and bayonets fixed. About 10 percent of his command was posted as sentries.

Meanwhile in Prophetstown, an assemblage of six hundred to seven hundred warriors, including Chippewa, Huron, Kickapoo, Muco, Ottawa, Piankeshaw, Potawatomi, Shawnee, Wyandot, and probably Winnebago, listened as the Prophet's inflammatory oratory filled the night. Here without a doubt were the seeds of Tecumseh's hoped-for confederation, but Tecumseh, the military leader, would never have sent his warriors into battle so capriciously. Shabonee, a Potawatomi chief, later testified that two Englishmen were present that evening in British red and that they urged the coalition to attack. Perhaps, but the Prophet had already lit the fuse. After his incantations—which included assurances that the enemy's gunpowder had turned to sand and his bullets to soft mud—the assembled warriors swarmed through the murky darkness to attack Harrison's camp.

Stephen Mars was the first of Harrison's sentries to hear movements through the thickets. He fired an alarm, but before he could reach the safety of his lines, he was struck down dead. This initial charge on the northwest quadrant of the camp overwhelmed a company of regulars and another of Kentucky volunteers. As they sprang to arms, many were silhouetted against their own campfires and made for easy targets.

Harrison was awake before the first shot and at its sound hurriedly pulled on his boots and called for his horse. There are many versions to the story, but the gist of it is that Harrison's gray mare was not immediately at hand, so he mounted the black horse of Major Waller Taylor. At the same time, Colonel Abraham Owen rode forward on a gray horse similar to Harrison's mare. Owen was immediately slain by a flurry of warriors who were lying in wait for the appearance of a gray horse, assuming its rider to be Harrison. When Waller Taylor mounted Harrison's gray mare and rode up beside Harrison, the general quickly told him to get another mount.

As the attack continued, Harrison ordered his men to hold their ground. In reality, there was little place for them to go, but cartridges loaded with number twelve buckshot added to Harrison's resolve. With dawn streaking the eastern sky, the warriors rushed forward in wave after wave against Harrison's defensive perimeter, and the Prophet's assurances of invincibility were soon proven false. Among those who fell on Harrison's side was

Kentuckian Joseph Hamilton Daviess. Clad in a white coat, Daviess made a prominent target as he attempted to lead a charge outside the perimeter. He would be mourned throughout Kentucky and remembered as the man who had prosecuted Aaron Burr at Frankfort after Burr's first indictment.

Now Harrison ordered his cavalry to harry the enemy's flanks, and soon the warriors were in disorganized retreat to Prophetstown. Harrison had held the field and the day. Records of casualties vary, in part because of those severely wounded in Harrison's command who later died en route back to Vincennes. Contemporary accounts reported 38 warriors slain on the field of battle with an unknown number of wounded, while Harrison lost 62 killed and 126 wounded.

A rumor circulated that Tecumseh was close at hand with reinforcements, but he was in fact hundreds of miles to the south. The next day Harrison found Prophetstown deserted, and he ordered it burned to the ground. Thanks to the Prophet's wild indiscretion, Harrison had accomplished his objective and made his enemy appear to be the aggressor. Despite this, Harrison's enemies were quick to brand him a shameless land-grabber and haphazard leader. His friends saw it more simply: he had saved the Northwest from annihilation. The truth, of course, lay somewhere in between, but the predawn skirmish that came to be called the Battle of Tippecanoe secured Harrison's reputation as a leader in the Northwest. The Prophet was not so lucky. He was the subject of almost universal scorn, although nothing would equal the disdain that Tecumseh would hold for his brother when he returned from the South to find both Prophetstown and the dreams of his confederacy in blackened ruins.[11]

Like much of the war to which it was a preface, the Battle of Tippecanoe was a relatively minor affair in military terms. One account of the battle called it "one of the most important conflicts which ever occurred between the Indians and the whites," and asserted that superior numbers of Indians were "entirely routed" due to the "gallantry, courage, and consummate generalship of Harrison."[12]

It was hardly that. But few, if any, battles of comparatively minor military significance have had such major political ramifications. After the Battle of Tippecanoe, they were twofold. In the short term, Harrison's destruction of Prophetstown drove Tecumseh and his followers irrevocably into the arms of the British. There was no longer any question about whose side Tecumseh was on. Harrison would meet him again among British troops along the Thames River in Ontario. Likewise, Tippecanoe inflamed American indignation that the British were behind the Indian attacks that escalated along the Northwest frontier in the aftermath of the battle. "The war on the Wabash is purely British," the *Lexington Reporter* editorialized within weeks of Tippecanoe. "The scalping knife and tomahawk of British savages, is now again devastating our frontiers."[13]

The long-term political ramification of the Battle of Tippecanoe was that almost thirty years later cries of "Tippecanoe and Tyler, too" would still reverberate and send William Henry Harrison to the White House. By then, he would be but a shell of the man who had once marched along the Wabash.

MR. MADISON'S WAR

So what were the British up to? Quite frankly, Great Britain had long had its hands full elsewhere than North America. To be sure, His Majesty's Government was keenly committed to retaining his Canadian possessions and equally desirous of curbing the march of the neophyte United States westward toward his lands in Oregon. But a key source of Great Britain's friction with the United States resulted indirectly from its larger conflicts with a diminutive corporal from Corsica who had ridden out of the ashes of the French Revolution and proclaimed himself emperor of France.

By 1812 Napoleon's military dominance on land had been almost unassailable for the better part of a decade. One after another, he had handily defeated the best armies that Great Britain's European allies could muster. Instead of reeling from Nelson's 1805 naval triumph at Trafalgar, Napoleon smashed an Austrian and Russian army at Austerlitz a few months later. Then in 1806 he beat the Prussians at Jena, and the Russians again the following spring at Friedland. Perhaps he should have been satisfied with the uneasy stalemate that followed—he master of a continent, Great Britain still mistress of the seas. But like other conquerors before and after him, Napoleon was intent on worldwide dominion.

Decisive battles gave way to complicated and indecisive commercial warfare. Napoleon fired the first broadside by issuing the

Berlin Decree in November 1806 and boldly declaring all commerce with Great Britain by any nation to be illegal. King George III's government retaliated with the Orders in Council, a series of edicts blockading all continental ports and barring all foreign ships from them unless those vessels first called at a British port and paid appropriate customs duties. Not that the United States or any other sovereign nation was about to pay duties to Britain for the right to trade with France, but in December 1807 Napoleon threw the final log on the smoldering fire. His Milan Decree declared that any vessel submitting to the British rules had in fact become British property and as such was subject to seizure by the French navy. Thus, whatever course United States ships took, they were apt to be at odds with either Great Britain or France.[1]

British impressment of American citizens was insult enough, but now American ships were being seized outright, sometimes by the Royal Navy and sometimes by France. Reacting in part to the anti-British fervor that swept the country in the aftermath of the *Chesapeake* affair and at the urging of President Jefferson, Congress passed the Embargo Act of 1807 even before news of Napoleon's Milan Decree reached America. The basic purpose of the Embargo Act was to prohibit *any* American exports. American vessels were not allowed to sail for any foreign port, and foreign ships were restricted from departing American ports with any cargo. Importing was not banned, but circumstances were rare when a captain called in an American port with a cargo, knowing that he could return to the high seas carrying nothing but ballast. Certainly, the Embargo Act would have been a death sentence to any nation—even when self-imposed—but it was particularly so to one that had just become so agitated over the acquisition of Louisiana in part to resolve its right of export through New Orleans.

However misguided, Jefferson's goals in promoting the Embargo Act were twofold. First, he hoped to resolve the nagging impressment and seizure controversies by simply removing American ships from the high seas. Well, he did that, all right. Second, he hoped to pressure both Great Britain and France into easing their arrogant and oppressive actions toward American shipping by denying them American goods.

Cutting off your nose to spite your face; throwing the baby out with the bathwater—characterize the Embargo Act as you will, it had a disastrous impact on the American economy. Exports plummeted from $108 million in 1807 to $22 million a year later; imports dropped almost as much—from $138 million to $57 million. The American economy almost choked to death. What kept it from doing so is that many American ships engaged in domestic coastal trade and then suddenly found themselves "blown off course" to foreign ports in the Caribbean. Many American ships abroad at the passage of the act simply stayed there, trading only between foreign ports and preferring to take their chances with British guns and French decrees rather than face certain economic ruin sitting bottled up in American harbors. Smuggling between Canada and the United States also mushroomed.[2]

In retrospect, the Embargo Act of 1807 stands as one of the most isolationist acts ever championed by an American president. How Thomas Jefferson managed to torpedo the American economy so savagely and still have his handpicked successor easily elected president in 1808 is less a commentary on the strength of his Democratic–Republican Party than on the weakness of the Federalist opposition. Needless to say, the dwindling number of Federalists representing the trade-oriented New England states were united in apoplectic rage against both the act and its principal architect.

The bitter pill of the Embargo Act was not abolished until the final week of Jefferson's presidency in 1809. In its place, Congress passed the Non-Intercourse Act, a watered-down version of the Embargo Act that forbade trade only with Great Britain and France and authorized the president to end those boycotts individually if and when either country stopped violating America's neutral rights by impressing seamen and seizing ships.

In truth, the Non-Intercourse Act wasn't much of an improvement. It proved equally difficult to enforce and most importantly seemed to have little influence on British or French policies. The United States might be acting in righteous indignation, but it was still the main party being hurt. By late in 1809, the measure was watered down further with legislation that kept American ports closed to Great Britain and France but permitted American ships to trade anywhere, including in the home ports of those two belligerents.

Trade with Great Britain rebounded somewhat, but trade with France remained severely hampered by the British blockade of the continent. Thus, the British continued to seize American ships that attempted to trade with France, and Napoleon continued to seize American ships and their cargoes wherever it suited him.[3] Clearly, the problems that had precipitated the Embargo Act remained unsolved, but by now they belonged to Jefferson's successor.

James Madison was Thomas Jefferson's philosophical and political heir. Yet another of Virginia's favored sons, Madison cut his teeth in the Continental Congress during the Revolution, played a key role in the Constitutional Convention, and finally followed the ascending road of the times by serving eight years as Jefferson's secretary of state. If ever a president handpicked a successor, it was Jefferson as he placed the mantle of presidential leadership on his fellow Virginian and used all his own political power to ensure Madison's election.

Short in stature and characterized by some as frail-looking, James Madison was a deep thinker of strong intellect. He was the type of person one would turn to for advice on the most complicated of matters, but not one expected to ride to the head of his troops and yell, "Charge!" Like Henry Clay's election as Speaker of the House two years later, Madison's electoral triumph was less one of personal popularity than one brought about by the declining power of the Federalists. Madison assumed the presidency in 1809 after a decisive win over Charles Pinckney, who had already carried the Federalist banner against Jefferson in 1804. Only New England stood solidly for Pinckney. The region's opposition to Madison would remain a thorn that Madison could never remove from his side and one that would come to have grave consequences for him as he tried to exert national leadership amid the growing crescendo of war.

New England, in fact, didn't like much of anything that was transpiring in the rest of the country. The region watched with disdain while Henry Clay and his fellow war hawks beat drums of war. Its key concern—and the spark plug of its economy—remained trade on the high seas. The Embargo Act and its progeny be damned. So what if a ship or crew was lost now and then to British high-handedness? It was a cost of doing business—a business that would be greatly disrupted further should Mr. Clay and his westerners launch all-out war. Indeed, the Federalists' contempt for Clay's saber rattling was matched only by their ambivalence toward Madison, particularly as it became increasingly evident that Madison lacked the charisma and force of personality either to rein in Clay and his war hawks or to rally the Federalists to the war cause.

Thus there was already considerable friction on all sides as factions across the country geared up for the presidential election of 1812. As an opening salvo, the incumbent tossed Congress what quickly became a political bombshell. On March 9, 1812, James Madison informed Congress of the existence of a British plot to encourage the New England states to secede from the Union—not, it should be noted, that they needed any encouragement!

As so often happens in politics, it was a crazy little affair that quickly took on a life of its own. At the center of it was an Irishman named John Henry. Born in 1777, Henry immigrated to the United States in 1798, and then made his way to Montreal to dabble in the fur trade. In 1808 he made a business trip to New England, and upon his return to Canada took it upon himself to send a report to the British government on the state of affairs that he had found there. The following year Sir James Craig, governor-general of Canada, was apparently impressed enough to commission Henry to make another trip to New England and secure additional information, particularly about Federalist opposition to any coming war.

This time on his return to Canada, Henry was paid 200 pounds (about $900) as compensation. The amount was a tidy sum for a "business trip" in those days, but Henry seems suddenly to have overestimated his importance as an informant. He spent the next two years in England and Canada, reportedly asking for everything from 32,000 pounds ($142,000) outright to an office paying 500 pounds a year. When no one paid him much attention—let alone British sterling—Henry fell in with a French ne'er-do-well passing himself off as the Count de Crillon. Incredibly, the impostor count suggested that Henry sell his reports to the United States government. Henry readily agreed.

Somehow, the two con artists secured a letter of introduction from Massachusetts governor Elbridge Gerry and wound up in Washington, where they persuaded James Madison's administration to buy the documents sight unseen for $50,000, then the entire budget of the State Department's clandestine operations fund. These events were made all the more ludicrous because Henry's accounts of Federalist dissatisfaction with the mounting calls for war were something that just about anyone could have read in Boston newspapers. Nonetheless, by the time Madison reported the transaction to Congress, he was convinced that Henry's papers contained "formal proof of the cooperation between the Eastern Junto & the British Cabinet," and that Henry was in fact a British agent dispatched to New England to stir the pot of secession and ultimately sever New England from the Union.[4] That quickly proved a vast overstatement, if not an outright fabrication.

Initially, the revelation of the Henry papers made Federalists in Congress look embarrassed at the least, treasonable at the worst. Closer examination, however, soon showed that there was nothing in Henry's papers to implicate anyone. When questioned on the subject of individual conspirators, Secretary of State James Monroe was forced to concede that the Madison administration was "not in possession of any names." "Where was Henry?" demanded the Federalists. "Out of the country," was the reply and oh, by the way, he had been promised immunity anyway.

Suddenly it was Madison and his Republicans, rather than the Federalists they sought to discredit, who were decidedly embarrassed over a $50,000 expenditure that netted a rehash of New England newspapers and barroom gossip. What the Henry affair did prove, however, was just how wide the rift was between the Madison administration and the New England Federalists. "We have made use of Henry's documents," Secretary of State Monroe confided to the French ambassador in an attempt to save some face, "as a last means of exciting the nation and Congress."[5] They did neither. Instead, most New Englanders would refer to the coming conflict as "Mr. Madison's war."

CONCESSIONS TOO LATE

So the United States had quite a list of grievances against its former sovereign: impressment of American sailors, provocation of Indian unrest on its frontiers, and the outright seizure of its commercial ships. Taken individually, each might have been enough to demand a course of war. Taken collectively, and fanned by Henry Clay and his Canada-hungry war hawks, to some Americans they most certainly were—no matter how militarily unprepared the United States might be.

And the United States was decidedly that. In January 1812 after considerable debate, Congress authorized increasing the regular army to 35,603 officers and enlisted men serving five-year terms, but that was far easier decreed than done. The effective strength of the United States Army was then only about 4,000 officers and men. A month later when Congress tried to bridge this gap between existing and planned size by empowering the president to call for 30,000 one-year volunteers, the question became whether the state militias that would likely provide them would be inclined to serve beyond the borders of their respective home states. The country was still that provincial.

In April, Clay's Congress authorized President Madison to offer an additional 15,000 eighteen month enlistments in the regular army and to require state governors to make ready as many as 80,000 volunteer militia to serve at a moment's notice. None of this happened quickly, of course, and by June 6, 1812, there were still only 6,744 officers and men serving in the regular army and another 5,000 or so in the newly authorized volunteer regiments.

The navy wasn't in much better shape. The U.S. fleet consisted of five frigates, three sloops, seven brigs, and an assorted collection of sixty-two coastal gunboats. These were crewed by some four thousand seamen, many of whom were still in their teens. When the Madison administration requested a dozen seventy-four-gun ships of the line and twenty new frigates, Congress merely authorized the repair of five laid-up frigates. The same Congress that urged expansion of the army voted down the creation of an oceangoing navy because of opposition from the war hawks—evidence to many that it was indeed cries of "Canada!" that quickened their pulses and not injustices on the high seas. Only the eighteen-hundred-man Marine Corps could boast of its relatively recent combat experience on the shores of Tripoli.[1]

It was hardly a state of affairs to inspire military confidence, no matter how upset the United States might be with tensions on the high seas and along its Canadian border. But the truth was that Great Britain was even more unprepared to fight a sustained war in North America because it had grown increasingly preoccupied with far more pressing concerns on the European continent. In the spring of 1812, Napoleon led his legions eastward into Russia and unleashed a series of events that in time would come to have as great an impact on the course of the ensuing American war as any military engagement in North America. If Russia fell to Napoleon's legions, Great Britain could well foresee the demise of its global empire and perhaps the world order. Even Thomas Jefferson had branded Napoleon "the first and chiefest apostle of the desolation of men and morals" and would come to call him "the Attila of the age."[2]

In that light, the issues causing friction between Great Britain and the United States appeared relatively small. There were many in England and America alike who agreed with a writer to the *London Times* that "the Alps and the Apennines of America are the

British Navy; if ever that should be removed, a short time will suffice to establish the head-quarters of a [French] Duke-Marshall at Washington . . ."[3] Indeed, those with any exposure at all to the situation in Louisiana knew well Napoleon's optimism toward regaining it. Thus, it seemed that France would be the clear beneficiary of any U.S. military action against Great Britain. If not with the direct intervention that Lafayette had once carried to the aid of the Thirteen Colonies, by declaring war on Great Britain the United States would nonetheless be coming indirectly to the aid of France.

Inconceivable as that seemed to some, there were those in Great Britain who knew that it had to offer some sort of olive branch to its disgruntled former colonies if for no other reason than to cover its flanks and prevent the dilution of its efforts against Napoleon. Accordingly, Parliament began to debate the repeal of the Orders in Council that would in effect normalize trade relations with the United States. But unfortunately, the War of 1812 was plagued by a dismal slowness in international communication both at its beginning and at its end. So while Parliament debated conciliatory concessions, the United States Congress debated a declaration of war.

The debate was unexcited and in many ways reluctant, but it mirrored the sentiments of the country as a whole. Sure, there were those young firebrands champing at the bit to ride for Canada and others ardently pushing the special interests they thought a war might benefit, but many in Congress, as in the country at large, failed to be galvanized by one clarion summons to war. The Henry papers fiasco had failed miserably to spark that summons and unite the country in moral outrage, and even on the most egregious of British trespasses there was no universal indignation for war.

Perhaps one of the most salient observations of this national reluctance came from the pen of Henry Adams three-quarters of a century later as he chronicled the administrations of Jefferson and Madison. "Many nations have gone to war in pure gayety of heart," Adams wrote, "but perhaps the United States were first to force themselves into a war they dreaded, in the hope that the war itself might create the spirit they lacked."[4]

Perhaps "spirit they lacked" was too harsh a judgment, but there was certainly a decided lack of enthusiasm that was widespread. In Massachusetts the Federalist-controlled legislature voted overwhelming against an "offensive war," counting many Republicans in their majority. Across the Hudson in New York, there was a growing antiwar sentiment despite, or perhaps because of, its proximity to Canada. In Virginia, crusty John Randolph and others in the western part of the state were opposed to war, as were certain congressmen in North Carolina.

Part of the national ambivalence may have been because even among those in favor of war, there was a nagging question about whom to fight. The Kentucky legislature railed against both Great Britain and France. Four other legislatures implied in resolutions that France was an enemy equal to Great Britain. Participants at a public meeting at Charleston, South Carolina, attended by Republicans and Federalists alike—including two-time Federalist standard bearer Charles Pinckney—complained bitterly that both Great Britain and France had provoked war on numerous occasions. In the midst of the war debate, there were calls for a "triangular war" against both Great Britain and France.[5]

Now here was more than a little Yankee brass! Great Britain and France were locked in an epic struggle for the control of a continent and dominion over half a world, and some in the United States boldly debated taking on both of them! There was actually ample provocation for this, despite the apparent political and military suicide. Later that summer, Madison sent a list of captured American ships to Congress and reported that since November 1807 the British had seized 389 vessels, but that the French had seized 558. Mollifying to some was the fact that the French did not impress sailors as did the British.[6]

Throughout the spring of 1812, Congress vacillated on the

war issue, at several points even coming close to adjourning until fall to delay confronting it. President Madison vacillated as well, time and again delaying sending a war message to Congress. Some contend he finally did so only to appease the war hawks in the western states whose electoral votes he would need in the fall, and in the hope that the declaration itself might loosen the jaw of the British lion. But in the end, after more than a decade of exasperation, to many it became a matter of national honor. "The period has now arrived," the Virginia House of Delegates resolved, "when peace, as we now have it, is disgraceful, and war is honorable."[7]

Thus, on June 1, 1812, Madison's war message was finally submitted and read before both houses of Congress. Given the custom of the day, it was read by a clerk who droned on without inflection or emotion for about half an hour. Not only was there no television, but both houses met in secret executive session. There were not even reporters in the galleries to play to. Even if there had been, an appearance before Congress would have placed Madison uncomfortably and uncharacteristically at the head of the charge.

Predictably, Madison's war message was devoted almost exclusively to maritime grievances: impressments, the Orders in Council, and ship seizures. Great Britain, the president charged, was determined to destroy American shipping and to secure for itself "the monopoly which she covets for her own commerce and navigation." He devoted only a paragraph to "the warfare just renewed by the savages on one of our extensive frontiers," but concluded that it was "difficult to account for this activity" without connecting it to the presence of British traders and garrisons with whom the Indians were "in constant intercourse." Great Britain, Madison asserted, was already in "a state of war against the United States." The decision before Congress was whether the country would "continue passive under these progressive usurpations," or respond with force and "commit a just cause into the hands of the Almighty Disposer of events. . . ."[8]

In Henry Clay's House of Representatives, the answer came

quickly, if less than decisively. Federalist opponents of the declaration first tried unsuccessfully three times to remove the veil of executive session and bring the debate into the public eye. Many congressmen had returned from spring visits home having found no groundswell of support for the measure. Still, many ultimately voted for it reluctantly and hoped, in the words of historian Bradford Perkins, that "the Senate would save the nation from the consequences of their own votes in the House." The result was that on June 4, 1812, the House of Representatives passed Madison's declaration of war 79 to 49. "I think," asserted New Hampshire's Josiah Bartlett, "the business was too hasty."

It was not as much of a sectional vote as New England opposition to Mr. Madison's war might have suggested. Clay's cohorts west of the Appalachians voted solidly for war, as did the delegations of Georgia and South Carolina, but the rest of the country was split. The Virginia and Pennsylvania delegations both voted a majority for war, but New York's was overwhelmingly opposed. Only Connecticut, Delaware, and Rhode Island cast unanimous antiwar votes, and in fact the measure may not have passed the House but for the prowar votes of six Massachusetts congressmen and the majorities of the Vermont and New Hampshire delegations.[9]

Over in the Senate, things moved more slowly. Perhaps the Senate would save the country from itself after all. After a week of debate, Alexander Gregg of Pennsylvania moved to send the war bill back to a select committee to amend it by substituting the declaration of full war with the issuance of letters of marque and reprisal. This would have authorized privateers to attack and seize British shipping, but, it was hoped, would limit the war to individual naval engagements. The Senate motion carried by a vote of 17 to 13, and more than one reluctant House member who had voted for full war breathed a sigh of relief.[10] Meanwhile, Vermont's Stephen Bradley was hastening south in a coach to add another vote against war.

But the sighs of relief were short-lived. Three days later, when the select committee reported out the marque and reprisal measure, the opponents of such a limited naval war defeated the proposal because one Republican changed his vote and two members who had not voted on the original Gregg motion now voted against it. The 16–16 tie meant that Madison's original war declaration was back on the table.

Then came tries at additional amendments, including one to issue letters of marque and reprisal against *both* French and British ships. It was defeated 15 to 17. With the defeat of each attempt at some measure of moderation, the inevitability of war appeared more certain. Reluctant though they might be, some senators began to consider changing their votes. Still, Stephen Bradley's coach hurried toward Washington.

James A. Bayard of Delaware, the same heartbeat who had once voted for Jefferson over Burr in the 1800 election, asked that consideration of the proposal be tabled until November. People weren't prepared for war, he said, and he wanted time for ships at sea to sail home safely. Bayard's amendment received only eleven votes and his subsequent try to delay a final vote until July garnered only nine votes. Finally, on June 17, 1812, the United States Senate voted 19 to 13 for the declaration of war.[11] It was by far and away the closest war vote in American history. James Madison was said to be "white as a sheet."[12] Perhaps he, too, had thought that the Senate would save him from the consequences of his decision.

It appears that "two or three senators shifted only when they saw that their negative votes would not prevent a declaration of war." Trying to put its most righteous spin on the vote, the Senate Foreign Relations Committee declared that "it must be evident to the impartial world that the contest which is now forced upon the United States is a contest for their sovereignty and independence." Two days after the vote, Stephen Bradley finally arrived from Vermont. History would ponder whether his earlier arrival might have held fast the reluctant few and changed the outcome.[13]

So the vote was done, the die cast, and James Madison signed the declaration of war on June 18, 1812, two days after the British

Parliament announced plans to repeal the Orders in Council and five days before it in fact did so. With its best eye on Napoleon's intentions, Great Britain assumed that such a concession would be enough to appease its American cousins. If news of this concession had reached the Senate before its final vote, might war have been averted? Probably not.

The Orders in Council and related trade issues were only one leg of the triangle of war sentiment. The British had shown no inclination to address or suspend their impressment practices, and the cries of the war hawks against the Indians and for Canada remained unabated. A British proposal for an armistice based on the repeal of the Orders in Council alone was flatly refused. Having managed to "let slip the dogs of war," Henry Clay was not about to leash them quickly. A Kentucky newspaper noted the armistice talk and described the reports that Canada might be left in British hands as "so ridiculous that we are almost ashamed to mention them."[14]

Clay and his war hawks looked around and breathed a sigh of relief. They had certainly not ridden a tidal wave of public opinion to this decision, but they had dodged the rocks in a river of rapids that had led irrevocably over the brink. The dogs of war were loose. The *Federal Republican* summed up the process for many when it editorialized that the country had been led kicking and screaming "by the blind and senseless animosity of a few 'new-hatched unfledged comrades,' who are but boys in public affairs, and who, in fact, have not been seen before by the American people on the public stage."[15]

Perhaps no one was more surprised at the declaration than the British ambassador, Augustus J. Foster. Summoned to the State Department by Secretary of State James Monroe on the afternoon after the vote, Foster reported to his government: "I have to remark on this extraordinary measure that it seems to have been unexpected by nearly the whole Nation; and to have been carried in opposition to the declared sentiments of many of those who voted for it, in the House of Representatives, as well as in the Senate, in which latter body there was known to have been at one time, a decided Majority against it."[16]

Monroe invited Foster to stay for tea, and Foster understood Monroe to hint that compromise might still be possible. Indeed, the Madison administration felt that any war would be limited in scope and/or brief in duration, and that perchance the mere declaration of war might be enough to force Great Britain to yield on the issue of impressment. Meanwhile, in the confusion of the next few months, at the very least Canada would likely drop like a rich plum into the American orbit. Reluctant warriors though they be, the Madison administration seemed confident of the outcome. Even before his meeting with Foster, Monroe wrote a friend: "My candid opinion is that we shall succeed in obtaining what it is important to obtain, and that we shall experience little annoyance or embarrassment in the effort."[17]

Events would soon prove, however, that it was difficult to engage in limited war. "By war," Elbridge Gerry of Massachusetts wrote President Madison, "we should be purified, as by fire."[18] And the fire was to come.

Lower Canada

Québec

Montreal

Maine (Mass.)

Crysler's Farm

Châteauguay

Upper Canada

Lake Champlain

VT (1791)

NH

Lake Superior

Fort Mackinac

Lake Huron

York

Lake Ontario

Sackets Harbor

USS *Chesapeake* vs HMS *Shannon*

Lake Michigan

Michigan Territory

Buffalo

New York

Boston

MA

Thames

Detroit

Lake Erie

Presque Isle

Hudson River

CT

RI

Fort Dearborn

Fort Meigs

Put-in Bay

Pennsylvania

New York

Ohio (1803)

Philadelphia

NJ

Indiana Territory

Ohio River

Pittsburgh

MD

Baltimore

DE

Washington

Potomac R.

Virginia

Chesapeake Bay

Kentucky (1792)

Mississippi River

Nashville

Tennessee River

Tennessee (1796)

North Carolina

South Carolina

Horseshoe Bend

Georgia

Charleston

Tombigbee River

Mississippi Territory

Alabama River

Savannah

Natchez

Fort Mims

Louisiana (1812)

Mobile

Pensacola

New Orleans

West Florida (Spain)

East Florida (Spain)

Apalachicola River

Gulf of Mexico

War of 1812
1812–1814

BOOK TWO
Bugles
(1812–1814)

The war, with all its vicissitudes, is illustrating the
capacity and the destiny of the United States to be a great,
a flourishing, and a powerful nation....

—*President James Madison to Congress, December 7, 1813*

OH, CANADA

Back in the heady days of 1810, war hawk Henry Clay brashly asserted to Congress that "the conquest of Canada is in your power. I trust that I shall not be deemed presumptuous when I state that I verily believe that the militia of Kentucky are alone competent to place Montreal and Upper Canada at your feet." By the spring of 1812, only four hundred Kentuckians had answered President Madison's first call to arms, and if Clay was having second thoughts, he kept them to himself.[1]

Still, on its face the matter of invading Canada seemed decidedly lopsided. In 1812 about 7.5 million people lived in the United States. Six million were white and the remainder African-American slaves. Canada's entire population was only about 500,000. The 100,000 or so residents of Nova Scotia and New Brunswick could be counted on to be strongly pro-British, but two-thirds of the 300,000 inhabitants of Lower Canada (modern Quebec) along the St. Lawrence River were of French descent. A full third of the remaining 100,000 inhabitants of Upper Canada (modern Ontario) north of the Great Lakes were American by either birth or descent.[2] How likely was it that these French and American populations would fight wholeheartedly for the British Empire? No wonder that after the declaration of war Governor Daniel Tompkins of New York confidently predicted that "one-half of the Militia of both provinces [of Canada] would join our standard."[3]

Only New England remained unimpressed. With most moderates rallying to the cause now that war had been declared, the *Connecticut Courant* put into print what was on the minds of many Federalists: The war "was commenced in folly, it is proposed to be carried on with madness, and (unless speedily terminated) will end in ruin."[4]

Meanwhile, there was considerable debate throughout the rest of the country about what to do with Canada once it was conquered. Should it be annexed to the Union and put on the path to statehood? (New England Federalists in the United States Senate were particularly ruffled at the thought of such further dilution of their influence.) Perhaps it would become a colony with lesser rights—although wasn't that what the previous war with Great Britain had been about? Then there were those who simply wanted Canada taken and held for ransom, a bargaining chip with which to force British concessions on the high seas. Within a week of the declaration of war, however, Secretary of State James Monroe conceded that public opinion might make it "difficult to relinquish Territory which had been conquered."[5] It was all bold talk—and decidedly premature. First, they had to conquer it.

For starters, the decision-making mechanism in place to conduct an offensive war of any sort was far from sound. The War Department consisted of Secretary of War William Eustis and eleven junior clerks. When Eustis complained about being overworked, Congress balked at authorizing two assistant secretaries, but it did reauthorize the quartermaster and commissary departments that it had abolished to save money back in 1802.[6] While the nominal chain of command ran from the president to Eustis and out to general officers in the field, there was no chief of staff or central command structure. To add further confusion, regular army officers routinely crossed paths with higher ranking militia officers of dubious military experience and debated whose authority was supreme over whose troops.

Even within this formula for chaos, however, it didn't take too much of a military strategist to recognize that Canada's lifeline of commerce was the St. Lawrence River and the Great Lakes. American commodore Isaac Chauncey spoke of conquering Canada by "taking and maintaining a position on the St. Lawrence" and thus "killing the tree" by strangulating its taproot.[7] Deprived of commerce, the argument ran, the tree of Upper Canada would wither and quickly look to the United States for sustenance. But as the Americans would come to learn, these waterways also afforded formidable natural defenses.

As Canada's largest city, Quebec on the St. Lawrence River was a logical target. But Quebec was also heavily defended, and there were still officers in the American army who personally remembered Benedict Arnold's failed campaign there during the Revolution. Besides, Quebec was due north of Federalist New England, and a campaign based from there might leave one wondering just where hostile territory began.

So in the end President Madison approved a plan championed by General Henry Dearborn, who had served as Jefferson's secretary of war. It proposed a three-pronged attack against Canada. One prong would be aimed at Montreal, which would avoid the problems of Quebec but still affect control of the upper St. Lawrence River. The other two prongs would attack the heart of Upper Canada through the logical gaps in the barrier of the Great Lakes: westward across the Niagara River between Lakes Ontario and Erie and eastward from Detroit at the western end of Lake Erie. The latter two routes had the added advantage of beginning deep in the heart of war hawk territory, expected to be a fertile source of volunteers.

In theory, the attacks along these three corridors should have been launched simultaneously. As it was, given the dismal state of military preparedness, the less than stellar acumen of the military leadership, and the confusions in communication and the chain of command, it was some wonder that they were launched at all.

· · ·

The western prong that was to attack east from Detroit was in fact already plodding toward there when war was declared. In the spring of 1812 three regiments of Ohio militia rendezvoused at Dayton, Ohio, and then marched north to join up with the regulars of the Fourth Infantry Regiment. Now under the command of Lieutenant Colonel James Miller, the Fourth Infantry had been at Vincennes since its return from Harrison's campaign to Tippecanoe. On June 10, 1812, at Urbana in west central Ohio, the Ohio militia greeted these veterans and marched with them through an arch proclaiming "Tippecanoe—Glory."[8]

To command this combined force of almost two thousand men, the Madison administration chose William Hull, originally of Massachusetts but of late the governor of Michigan Territory. Now fifty-nine, Hull had led Massachusetts troops with some distinction during the Revolution. He had also been the nominal head of Michigan's territorial militia since his appointment as the territory's first governor in 1805. Militarily, there was not much else to recommend him. Genial by nature, Hull appears to have lobbied for the appointment as a brigadier general, but in the aftermath of the campaign chose to remember that he had accepted it "with great reluctance."[9]

To his credit, Hull warned the War Department that control of Lake Erie was essential to any successful military operation near Detroit. Yet having said that, Hull was somehow persuaded otherwise, and with British ships sailing Lake Erie at will, Hull marched overland toward Detroit. Just how isolated Michigan Territory still was is evidenced by the fact that Hull's first task was to carve a road northwest from Urbana to the falls of the Maumee River to facilitate a line of supplies for his army. Once on the Maumee, Hull made his first tactical mistake.[10]

The schooner *Cuyahoga* was on the Maumee and appeared to offer a quick and easy passage to Detroit. Onto the *Cuyahoga*'s decks Hull ordered a complement of sick soldiers, some supplies, and a trunk containing not only the detailed muster rolls of his army, but also complete copies of his correspondence with the

War Department. Later, when recriminations blew stronger than a Lake Erie gale, there would be conflicting stories about whether Hull himself had ordered the trunk aboard or whether such had been the work of his aide, Captain Abraham F. Hull, who also happened to be his son.

Regardless of the responsibility for the trunk, Hull ignored repeated warnings that the *Cuyahoga* was sailing into trouble. To reach Detroit, the schooner had to pass under the guns of the British post at Amherstburg, also called Fort Malden. True, no news of a formal declaration of war had yet reached either side, but with all reports suggesting that war was imminent, it was doubtful that the British would permit what was clearly a resupply ship to pass up the Detroit River unmolested.[11]

As the *Cuyahoga* sailed off, Hull's main force continued through the Black Swamp of the Maumee, crossed the River Raisin, and arrived at the gates of Detroit on July 5, 1812. By then, Hull had proof positive that a state of war existed with Great Britain—both from dispatches that had caught up with him and from the British capture of the *Cuyahoga*, just as many had predicted. "I had no idea," reported British major general Isaac Brock upon learning of the contents of Hull's captured trunk, "that General Hull was advancing with so large a force." Now Brock, who had been appointed governor of Upper Canada and commander of its armed forces the previous year, had full details about his enemies and the ultimate objectives of their advance to Detroit.[12]

Founded in 1701 by Frenchman Antoine de la Mothe Cadillac on the short river linking Lake St. Clair and Lake Erie, Detroit—a French word meaning "strait"—was by 1812 a village of about 160 homes and seven hundred inhabitants. Fort Detroit sat on a rise above the town and about 250 yards back from the river. It was not an ideal location from which to command the river or fire on the opposite Canadian shore because any artillery fire had to be lobbed over the town. Lacking though he was in some military matters, Hull made up for them upon his arrival by issuing a flowery general order announcing both his assumption of command and Congress's declaration of war. A week later, Miller's Fourth Infantry Regiment and one regiment of Ohio militia rowed across the Detroit River and landed on the Canadian shore—unopposed.[13]

The Americans were giddy with success. Hull's aide-de-camp, Lieutenant Robert Wallace, wrote from occupied Sandwich that "the British cause is very low in this province and their militia and Indians are deserting by hundreds. Our Flag looks extremely well on his majesty's domain."[14] Hull, more the politician than the soldier, got caught up in the excitement himself and issued a pompous proclamation. "Inhabitants of Canada! You will be emancipated from tyranny and oppression," Hull promised, "and restored to the dignified station of freemen. Had I any doubt of eventual success I might ask your assistance, but I do not. I come prepared for every contingency."[15] With that Hull turned south and marched down the eastern bank of the Detroit River to lay siege to Fort Malden.

But chinks rapidly developed in Hull's boastful position. To begin with, he dallied in front of Fort Malden, engaging in several skirmishes but failing to take the fort even though his force outnumbered its occupants at least two to one. Meanwhile, two hundred of his Ohio militiamen set what would become an all too common precedent and refused to cross the river and serve outside American territory. The Canadian militia and their Indian allies—far from deserting by the hundreds—had simply melted into the woodlands to await a coordinated attack by General Brock. And despite Hull's liberating proclamation, the citizens of Upper Canada had generally failed to flock to his banner.[16]

Hull looked around and suddenly became very uncomfortable about his lines of supply running back to Ohio. On the east they were threatened by British ships on Lake Erie, just as he had predicted, and on the west by Indians still smarting over Harrison's treatment at Tippecanoe. When Hull dispatched 150 men under Major Thomas Van Horne to meet a supply train from Ohio at the River Raisin just thirty-five miles south of Detroit, the detachment was attacked by warriors led by Tecumseh and beaten back to Detroit. Next, Hull dispatched a larger force of 600 regulars and militia under Colonel Miller to break through, but this force, too, was forced to retreat to Detroit.[17]

Then came word that left Hull feeling really surrounded. The American outpost on Mackinac Island had surrendered to a combined force of Indians and British. Called Michilimackinac by both the French and the British, the island guarded the straits between Lake Huron and Lake Michigan and was the crossroads of the upper Great Lakes. Some said that Michilimackinac was a Chippewa word meaning "green turtle" and that the two-mile-wide and three-mile-long island was so named because of its humped oval shape and dense covering of dark pine forests. Colloquially, the Americans had shortened the name to Mackinac and pronounced it "Mackinaw."

In 1670 French Jesuit Claude Dablon erected a crude mission on the island, although a year later Father Jacques Marquette moved it to the northern shores of the strait and renamed it St. Ignace. About 1714 the French hacked wooden Fort Michilimackinac out of the dense forest on the southern shore just west of where the massive suspension bridge now spans the straits and links Michigan's upper and lower peninsulas. In the ensuing decades, everyone who was anyone in the Northwest fur trade passed this way.

On a warm June morning in 1763, the fort was under British control. A group of seemingly peaceful Chippewa gathered outside and proceeded to play what the Canadians called lacrosse. As the British soldiers slowly ambled outside the stockade to watch, squaws with weapons hidden under their blankets casually slipped through the gates. When the leather ball flew through the gates and into the fort as if by accident, the Chippewa followed the play. Once inside, they grabbed the hidden weapons and proceeded to slaughter nearly every British soldier.[18]

Fort Michilimackinac was again in British hands during the

American Revolution. After frontiersman George Rogers Clark caused alarm by capturing British posts at Kaskaskia, Cahokia, and Vincennes in the Mississippi Valley, the British commander at Michilimackinac, Patrick Sinclair, determined to remove his garrison to Mackinac Island and a more secure position. Sinclair proceeded to construct an elaborate stone bastion that melded into the island's hillside and commanded both the harbor below and the straits beyond. Impregnable as the new fort seemed from the water, however, it was surprisingly vulnerable to any enemy guns that might be placed above it higher on the hillside. Sinclair shrugged off this weakness, but thirty years later the British themselves would remember it.

The Treaty of Paris of 1783 ending the Revolution decreed that all British posts along the lakes on the American side be surrendered to the United States "with all convenient speed." That proved a relative term, and it was only after Jay's Treaty that the Americans finally took possession of Fort Mackinac, late in the fall of 1796. The British established a new post some fifty miles to the east on St. Joseph Island between Lake Superior and Lake Huron. The village below Fort Mackinac, however, remained decidedly British and French, and its inhabitants continued to sit astride a lucrative fur trade with scarcely a glance to their new landlord.[19]

In the summer of 1812 the American commander at Fort Mackinac was Lieutenant Porter Hanks. His opposite number at Fort St. Joseph was Captain Charles Roberts. When a French-Canadian trader named Toussaint Pothier heard in Montreal of the declaration of war, he paddled up the Ottawa and across most of Upper Canada to deliver the news to Roberts. Now, Roberts had one thing that Hanks lacked—the element of surprise. Incredibly, despite being governor of Michigan Territory, General Hull had failed to inform Hanks, and arguably the most important command in his territory outside Detroit, that the country was at war with Great Britain.

Pothier and other traders urged Roberts to attack Mackinac

immediately. Roberts made preparations to do so, but took no action until he received formal orders from Brock instructing him to take the "most prompt and effectual measures to possess himself of Michilimackinac," or in the alternative to defend St. Joseph most aggressively in the event of an American attack. Brock advised Roberts to make the most of friendly Indians and fur traders in doing either. Thus, on the afternoon of July 16, 1812, Roberts mustered only 46 regulars but bolstered them with about 180 Canadian militia and traders and some 400 Indians, mostly Chippewa and Ottawa, and sailed west in a collection of canoes, barges, and the brig *Caledonia*.[20]

Back on Mackinac Island, Hanks and his garrison of fifty-seven regulars were ignorant of the state of war, but not completely clueless to events. Why were traders and Indians not coming to the island in their usual numbers? Why did they seem to be congregating to the east at St. Joseph? When Hanks sent a trader named Michael Dousman east to St. Joseph to investigate, Dousman's canoe paddled smack into the middle of the British invasion flotilla. Dousman was captured and returned to the island in the wee hours of July 17. He was ordered to awaken the villagers and assemble them at a point away from the fort. Meanwhile, the bulk of Roberts's force landed in a small cove on the northwest coast almost directly across the island from the fort and proceeded to haul a six-pound cannon up the heights behind and above it.

Hanks was finally alerted to all this by the village doctor, who managed to elude the round-up party and make his way to the fort. Hanks briskly turned out his troops and made preparations for the post's defense, but he was outnumbered and soon outgunned. By 9:00 A.M. Hanks looked up the hillside above his position to the six-pounder staring down at him. Perhaps he thought of Captain Sinclair. Two hours later, Toussaint Pothier approached the gates under a flag of truce from Captain Roberts. The message was simple: "surrender or face the uncertain actions of our Indian allies." As Hanks later reported to Hull, "This sir, was the first intimation I had of the declaration of war." Hanks consulted with his officers and promptly surrendered without firing a shot.[21]

When Hull received word on July 28 that Fort Mackinac had surrendered, he exclaimed, "The whole northern hordes of Indians will be let loose upon us."[22] Unfortunately, Hull actually helped that scenario come true by immediately ordering the evacuation of Fort Dearborn on Lake Michigan. Originally built in 1803 at the site of what became Chicago, the log fort was hardly an impregnable bastion, but in the summer of 1812 it was well stocked with supplies and manned by a small garrison of fifty-some regulars under the command of Captain Nathan Heald. How long the fort might have held out is debatable, but Hull's evacuation orders sent many of its occupants to certain doom.

Early on the morning of August 15, 1812, Captain Heald reluctantly marched out of Fort Dearborn with his command of fifty-four regulars, twelve militia, nine women—including his wife, Rebekah—and eighteen children. Supposedly, they were to be escorted to the safety of Fort Wayne by trader William Wells and a group of about thirty Miami warriors. Wells had been captured by the Miami as a child and later married a sister of the great Miami chief Little Turtle. Despite this, Wells had left the Miami and fought against them at Fallen Timbers. Little Turtle had forgiven him. Tecumseh, who was also at Fallen Timbers, had not.

From near Detroit, Tecumseh was in contact with about five hundred Potawatomi and Winnebago warriors in the vicinity of Fort Dearborn. Hull had ordered Heald to distribute certain supplies to them as part of his evacuation. This hardly appeased those warriors who had been at Tippecanoe, especially since Heald destroyed the stocks of firearms and barrels of whiskey that they really coveted. They needed little encouragement from Tecumseh to attack.

Amid some sand dunes scarcely more than a mile from the fort, the Potawatomi and Winnebago fell upon Heald's column. The captain tried to organize a retreat but was quickly surrounded by overwhelming odds. The fight soon proved hopeless, and Heald approached the Potawatomi chief, Black Bird, to surrender. Black Bird accepted, but did little to stop the indiscriminate slaughter that continued, including that of two women and twelve children. Tecumseh got his revenge when William Wells was beheaded and his heart eaten raw. Captain Heald was luckier. Though both were badly wounded, he and his wife were among the few survivors. They were captured briefly and then managed to make their way to Fort Michilimackinac, where British captain Charles Roberts proved a gracious, if unexpected host.[23]

Meanwhile, the man who had ordered Heald to evacuate was having evacuation thoughts of his own. In light of reports that General Brock was advancing with an unknown number of reinforcements, Hull withdrew his troops back across the Detroit River. Such action was not well-received by the regimental commanders of the Ohio militia. Colonels Duncan McArthur, Lewis Cass, and James Findley quickly lost what little confidence they had ever reposed in Hull and solicited Colonel Miller of the regular Fourth Infantry to take command. Regular officer that he was, Miller undoubtedly understood the meaning of the term "mutiny" and politely refused their request, no matter how uncomfortable he might be with Hull.

Lewis Cass, later Michigan territorial governor himself, wrote to Ohio governor Return Jonathan Meigs to plead their case: "From causes not fit to be put on paper, but which I trust I shall live to communicate to you, this army has been reduced to a critical and alarming situation." Cass urged Governor Meigs to send at least two thousand reinforcements and do all that he could to keep open the lines of supply and communication between Ohio and Detroit.[24] In point of fact, of course, what remaining men and supplies Meigs could assemble—as Hull and the colonels knew only too well—had been unable to reach Detroit and were bogged down at the River Raisin. (Half a century later at the outbreak of another war on American soil, Lewis Cass would resign as James Buchanan's secretary of state in protest over the president's failure to reinforce Fort Sumter.)

By the way, Return Jonathan Meigs deserves mention if for no other reason than his interesting name. Family legend has it that his grandfather was courting a girl in Connecticut when she rejected his proposal. Dejected, he began to walk down the lane, only to have the girl come running after him shouting, "Return, Jonathan!" They married after all, named their firstborn after those words, and the name was passed down to their grandson.[25]

With the Americans in retreat, the British lost no time in bringing three cannon and two mortars to bear on Detroit from the Canadian side. Brock indeed arrived with some three hundred reinforcements, having sailed along the northern shore of Lake Erie. He proceeded to make the most of Hull's timidity. Boldly, Brock sent a message to Hull on August 15 demanding the surrender of the American fort. "It is far from my intention to join in a war of extermination," wrote the proper British officer, "but you must be aware, that the numerous body of Indians who have attached themselves to my troops, will be beyond control the moment the contest commences."[26]

Hull refused the demand, but was clearly rattled. Seemingly oblivious to reports that Brock had landed troops on the American side, he slunk inside the walls of the fort and cringed before the intermittent cannon fire that continued to rain in from the Canadian shore. He was doing just that the next day when a cannonball came crashing into the officers' mess. Among the four killed was the hapless Lieutenant Porter Hanks, who was awaiting a court of inquiry over his recent surrender of Fort Mackinac.[27]

That was enough for Hull. He sent his son across the river under a flag of truce to accept Brock's terms and promptly surrendered his entire army. "Not an officer was consulted," reported Lewis Cass, and "even the women were indignant at so shameful a degradation of the American character." Ordered into the fort to stack arms, James Findley's men expressed their sentiments more pointedly: "Damn such a general." At noon on August 16, 1812, with General Brock at its head, a British column marched into Fort Detroit.[28]

Later, Hull would say that he had been short of ammunition, but clearly he had been short of courage. Returned to the United States on parole, Hull was later court-martialed and convicted of cowardice and neglect of duty. He was sentenced to death but then spared because the court considered his prior service during the Revolution and his advanced age. Hull would go to his grave in 1825 claiming that he had been the victim of the War Department's general ineptitude and a failure to coordinate the three northern attacks. That Brock was able to show up below Fort Detroit rather than be kept busy along the Niagara lends some credence to this claim, but it is still difficult to reconcile with the undisputed fact that Hull surrendered his command with what one account termed "unsoldierly alacrity." With Detroit in British hands, Brock hurried east to parry the American thrust at Niagara.[29]

The loss of Mackinac, Dearborn, and Detroit was a triple blow and left the entire sweep of the Northwest frontier open to attack by Indians and British. The only person who staved off a complete rout all the way back to the Ohio River was a young captain named Zachary Taylor, who commanded a spirited defense of Fort Harrison against an Indian attack on the night of September 4–5, 1812. More than a few people were left wondering where Clay's vaunted militia was about now.

The Madison administration was incensed and looked to replace the disgraced Hull with another regular army officer, but folks in the Northwest would have none of that. To save the day they once more turned to one of their own, the hero of Tippecanoe, Indiana governor William Henry Harrison. Kentucky made Harrison a major general of its militia even though he was not a citizen of the state, and Kentucky congressman Richard Mentor Johnson led the charge to give Harrison command of the entire Northwest. Madison finally agreed. Harrison received orders from Secretary of War Eustis "to regain the ground that has been lost by the Surrender of Detroit & the army under General Hull, and to prosecute with increased Vigor the important objects of the Campaign."[30]

Throughout the fall of 1812, Harrison went about assembling and training his force in the same methodical way that he had employed before the march up the Wabash to Tippecanoe. This time, the march would be all of the way to Detroit. To open the counterattack, Harrison dispatched an advance guard of about 850 under Brigadier General James Winchester to reconnoiter as far as the old Fallen Timbers battleground at the falls of the Maumee. Winchester should have stopped there, but he took it upon his own initiative to advance to the River Raisin.

Winchester's force was no luckier there than some of Hull's troops had been, and the column was attacked by about eleven hundred British and Indians on January 21, 1813. About one-third of Winchester's command was killed, and as at Fort Dearborn, surrender terms were ignored and many murdered after surrendering. "The savages were suffered to commit every depredation upon our wounded," read one report. "Many were tomahawked, and many were burned alive in the houses."[31] Now the Northwest had another battle cry to place less gloriously beside "Tippecanoe." "Remember the Raisin" would reverberate in Harrison's coming campaign the following year.

Meanwhile, the two eastern prongs of the 1812 offensive had not fared any better. To lead the advance across the Niagara frontier, the War Department approved New York governor Daniel Tompkins's choice of Stephen Van Rensselaer, a political appointment if there ever was one. Van Rensselaer was nominally a major general of volunteers, but he had no prior military experience. His credentials were that he was a member of the Federalist elite, and Tompkins hoped that his appointment might persuade other Federalists to become more enthusiastic about the war effort. For sound military advice, the reasoning ran, Van Rensselaer could

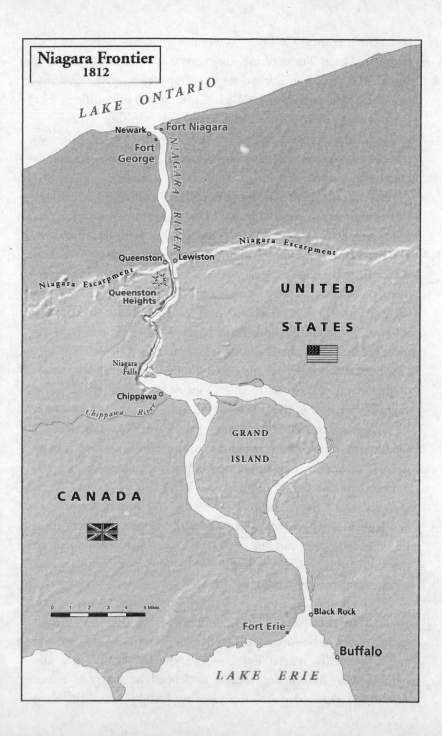

Niagara Frontier
1812

LAKE ONTARIO

Newark ○ ■ Fort Niagara

Fort
George

NIAGARA RIVER

Niagara Escarpment

Queenston ○ ○ Lewiston

Niagara Escarpment

Queenston
Heights

UNITED

STATES

Niagara
Falls

Chippawa ○

Chippawa River

GRAND

ISLAND

CANADA

0 1 2 3 4 5 Miles

Black Rock

Fort Erie

Buffalo

LAKE ERIE

rely on his cousin, Colonel Solomon Van Rensselaer, who had fought at Fallen Timbers and long served as New York's adjutant general. General Van Rensselaer was supposed to share his command in western New York with a regular army officer, but Brigadier General Alexander Smyth, whose own field experience was negligible, refused to obey the War Department's orders to place his force at Van Rensselaer's disposal. It was going to be that kind of campaign.[32]

By October 1812 Van Rensselaer, with no help from Smyth, had more than thirty-five hundred troops facing a force of two thousand British and Indians along the Niagara River at Lewiston. Van Rensselaer had hoped to persuade Smyth to launch a coordinated attack against Fort George six miles to the north near the river's mouth on Lake Ontario—downriver as the Niagara flows. Smyth declined, and Van Rensselaer determined to seize Queenston Heights directly across from Lewiston on his own. Here the river had high banks but was only about 250 yards wide. What followed was indicative of the disorganization, muddled leadership, and ineffective chain of command that was to characterize most of the battles of the war.

General Van Rensselaer ordered troops to cross the river on October 11, only to find that all the oars for the boats had "disappeared" downriver in a single boat. Was it treachery or just stupidity? Two days later, Solomon Van Rensselaer led two hundred men across the river, only to be wounded six times in the process and have his men pinned down under the heights. Captain John E. Wool took over and began to lead his company of regulars of the Thirteen U.S. Infantry Regiment to the top of the heights via an unguarded fisherman's trail.

Half a dozen miles to the north at Fort George, General Isaac Brock heard the sound of the opening American guns and paused momentarily to be sure that the attack was coming at Queenston and not directly against Fort George. Finally satisfied that the heights were indeed the target, he leaped to his gray horse, Alfred, and galloped south alone, leaving even his two aides in the dust. Reaching Queenston, Brock rode immediately up the heights to the British redoubt that was perched above the river and from which elements of the Forty-ninth Regiment were pouring a withering fire upon the Americans below.

Suddenly Captain Wool's company came streaming down on the redoubt from the crest of the heights. There was barely time for the British to spike their single cannon and then beat a hasty retreat into the village below to rally more troops. Brock could have given any number of orders, delegated the counterattack to anyone else. Instead he gathered about one hundred men of the Forty-ninth Regiment and a like number of militiamen and led an immediate charge to retake the redoubt, surmising correctly that if Wool's position was reinforced, the day would be lost.

Up the slope they went. In cocked hat, red coat with golden epaulettes, and decorative scarf given him by Tecumseh, Brock must have known that he made a prominent target. The first bullet merely pierced his wrist, and he barely paused. A second caught him square in the left breast, and he fell. Now for the stuff of legend. Some say that Brock was dead before he hit the ground. Others insist that with his dying words he urged those near him to "Push on the York Volunteers." Regardless, the York militia did come on, sweeping up the hill in a counterattack and momentarily recapturing the redoubt.[35]

To support Wool's efforts to retake it for the Americans, General Van Rensselaer turned to a regular army lieutenant colonel named Winfield Scott, who had just refused to waive his rank and serve under Colonel Solomon Van Rensselaer in the initial assault. The twenty-six-year-old Scott had already been in more than a little hot water. Born in Virginia and trained as a lawyer, Scott had been in the Richmond courthouse watching Aaron Burr's trial in the summer of 1807 when news arrived of the *Leopard*'s attack on the *Chesapeake*. He immediately sought a position in the Virginia militia. By the following spring, Scott was commissioned a captain in the regular army and soon ordered to take a company of artillery to New Orleans. At the time the commanding officer there, and indeed still the ranking officer in the United States Army, was none other than Brigadier General James Wilkinson. Scott had observed Wilkinson at length during the Burr trial and was not impressed. Unfortunately for him, Scott failed to keep his own counsel and aired his views on the general while in uniform, going so far as to announce that "he never saw but two traitors, General Wilkinson and Burr" and that Wilkinson was a "liar and a scoundrel."[34]

Accurate though that assessment might have been, it was not an appropriate characterization for a junior officer to make of his superior. Scott was hauled before a court-martial on charges of "conduct unbecoming an officer and gentleman" and a trumped-up charge of embezzling certain clothing allowances. The court recognized the trumped-up charges for what they were and dismissed them. It then chose to convict Scott only of "unofficer-like conduct"—a lesser degree than "conduct unbecoming." He was sentenced to a year's suspension, in no small measure because many on the court secretly applauded his denunciations of Wilkinson. Scott served his year in exile practicing law, and when war was declared he was suddenly vaulted to the rank of lieutenant colonel and placed second in command of the Second Artillery. When the smoke and fury atop Queenston Heights told of Wool's fight, Scott received General Van Rennsselaer's permission to hasten across the Niagara and take command.[35]

By the time Scott arrived on the heights, Wool's men had succeeded in recapturing the redoubt. Scott attempted to unspike the cannon that Brock had stood beside an hour before. Soon he had six hundred men on the heights, but Brock's second-in-command, Major General Roger Hale Sheaffe, was advancing from Fort George with its garrison, about one hundred Mohawk, and more elements of the soon-to-be storied York militia. Scott's position had to be made secure.

Confident now of victory, General Van Rennselaer ordered the New York militia to cross the river and reinforce Scott. They refused to leave American soil. "To my utter astonishment," reported a perplexed Van Rensselaer, "I found that at the very moment when complete victory was in our hands, the Ardor of the unengaged Troops had entirely subsided. I rode in all directions" and "urged men by every Consideration to pass over, but in vain."[36]

Now Scott was on his own. General Van Rensselaer sent him a message to evacuate his position and promised boats to move his command back across the river. In the face of Sheaffe's advance, Scott withdrew to the river, only to find no boats. With the British once again in command of the heights above, Scott had little choice but to surrender his command. In all the Americans lost more than 300 killed and wounded and had 958 taken prisoner, including Winfield Scott. British losses were 14 killed, 77 wounded, and 21 missing. Sheaffe received a baronetcy for saving the day at Queenston, but it was Brock who had given his life to permit him the opportunity. Brock had been made a Knight of the Bath for his capture of Detroit, but he died below Queenston Heights without knowing of the honor.[37]

Madison's Republicans were aghast at the defeat and quick to blame Van Rensselaer, some even going so far as to suggest that the Federalist general had warned the British prior to the attack. Van Rensselaer was truly distressed at the outcome and asked to be relieved of his duties. The War Department obliged, but then made the mistake of placing General Smyth in command. "Van Bladder," as his troops derogatorily called him, decided to shift the attack upriver to Fort Erie where the lake emptied into the Niagara River. Smyth proved long on boasts but short on action, and even when he managed to cross the river and secure positions below the fort, Smyth's officers voted down further action because this time it was the Pennsylvania militia that refused to cross the border. Smyth was relieved of command and unceremoniously dropped from army rolls.[38]

So the Niagara thrust had gotten nowhere. Little had been accomplished save perhaps the loss of the noble Brock, and Winfield Scott's first test in battle. Farther to the east, what should have been the Americans' key thrust of the three-pronged attack proved stillborn. The plan's architect had been sixty-one-year-old Henry Dearborn, yet another Revolutionary War icon, who had first marched to Quebec with Arnold in 1775. Known as "Granny" to his troops, he was as uninspiring a leader as Hull and Smyth had proven. His original plan had been to coordinate his attack against Montreal with that of Van Rensselaer and Smyth, and in so doing take pressure off Hull in the west. But Dearborn dallied in New England trying to raise troops. By the time that an exasperated Secretary of War Eustis finally ordered him to go to Albany or Lake Champlain and strike a blow toward Montreal before Congress met, Brock had successfully dispatched Hull and lost his life routing Van Rensselaer and Smyth. Dearborn's assault got no farther than the Canadian border, where once again state militia refused to cross into Canada. The army soon retreated, and a contemporary described its foray as a "miscarriage, without even the heroism of disaster."[39]

There it was. On land, the American campaigns of 1812 had been unmitigated disasters. Even Henry Clay had changed his tune. "Canada was not the end but the means," the Speaker now asserted, "the object of the War being the redress of injuries, and Canada being the instrument by which that redress was to be obtained."[40] But even that was now far easier said than done. The Northwest was in danger of collapse, there was stalemate at Niagara, and the assault toward Montreal had amounted to less than a pinprick.

Perhaps remembering Monroe's assurance of an easy victory, Secretary of the Treasury Albert Gallatin acknowledged that "the series of misfortunes exceeds all anticipations made even by those who had least confidence in our inexperienced officers and undisciplined men."[41] And far from welcoming the Americans as liberators, the Canadians had shown that they were unwilling to rise up against their British landlords. The first cries of "Oh, Canada" had ended on a sour note.

HURRAH FOR OLD IRONSIDES

When the grim news of Hull's surrender at Detroit reached Washington in late August 1812, the Madison administration looked around for some good news. It found it on the high seas that had borne so many of the causes of the present conflict. By some coincidence, the hero had the same name as the disgraced general— Hull. By even greater coincidence, he was the general's nephew.

Isaac Hull was born on March 9, 1773, at Derby, Connecticut, a little town on the Housatonic River at the head of tidewater some dozen miles from Long Island Sound. His father, Joseph, was William Hull's brother, and Isaac was the second of Joseph's seven sons. Like his brother, Joseph Hull had fought in the Revolution, leading a flotilla of whaleboats that harried British movements on Long Island Sound and even managed to capture an armed schooner. Thus, the sea was in young Isaac's blood, and on March 13, 1798, he was commissioned a lieutenant in the United States Navy and assigned to the USS *Constitution*.[1]

When Lieutenant Hull stepped aboard her decks, the *Constitution* was less than a year old and as untried in battle as he was. Built in Boston at the cost of some $300,000, the frigate was 204 feet long, with a beam of 43 feet. That was not a very big space in which to hold a crew that usually numbered upward of four hundred. But such numbers were necessary both to man the *Constitution*'s standard complement of forty-four guns and to

work the seventy-two different sails that adorned her three principal masts, including a main mast that rose 233 feet. When all
her sails were set, the *Constitution* carried some forty-three thousand square feet of canvas—enough to cover an acre—and in a
good wind was capable of making better than twelve knots.[2]

The *Constitution* was the third of three large frigates built to
create the fledgling United States Navy and attempt to project
American power on the high seas. Lieutenant Hull's first action
aboard her was during an undeclared naval war with France. By
its conclusion in 1801, Hull had served as the *Constitution*'s first
lieutenant (executive officer) and at the heady age of twenty-eight
had been temporarily appointed her captain. When things heated
up with the Barbary pirates, Hull expected to sail again for the
Mediterranean at the helm of the *Constitution*, but instead he saw
action there aboard the *Adams*, *Enterprise*, and *Argus*. After HMS
Leopard caught James Barron with his gun ports closed on the
USS *Chesapeake*, Hull was one of those who served on the court of
inquiry that found Barron accountable. Later, Hull himself commanded the *Chesapeake* before once more being given command
of the *Constitution* on June 17, 1810.[3]

The finer points of the *Constitution* and the classifications and
armaments of early nineteenth-century sailing ships could fill many
pages and provide grist for countless hours of roundtable debate
among naval historians. Indeed, Theodore Roosevelt in his seminal
The Naval War of 1812 discussed in exhausting detail the number of
guns and their respective poundage that opposing ships carried.
With all due respect to TR, here is the matter simply put.

Sailing ships in 1812 were classified first by their type (sloop,
brig, frigate, or man-of-war—the last sometimes called a ship of
the line) and then rated by the number of their *standard* complement of guns. Generally, a sloop carried one mast and one deck; a
brig, two masts and one deck; a frigate, three masts and a main gun
deck and upper, or spar, deck; and a man-of-war, multiple masts
and decks. The most famous of the American ships, *Constitution*,

United States, and *President*, were rated as forty-four-gun frigates. Where the finer points of discussion really come in is that captains frequently crammed a few more cannon onto their decks. Hence, the forty-four-gun *Constitution* carried fifty-two guns when she met the thirty-eight-gun frigate HMS *Java* that herself was overloaded with forty-nine guns.

Why was this important? In two words: broadside weight. The question was simple. How much weight could be slammed into an opponent at one time? On a forty-four-gun frigate with twenty-four 24-pound long guns and twenty carronades throwing 32-pound shot, the math for one side of the ship firing a broadside in unison was twelve long guns firing 24 pounds each equal 288 pounds, plus ten carronades firing 32 pounds each equal 320 pounds for a total broadside weight of 608 pounds. Roll your eyes if you will, but aboard creaking and groaning wooden sailing ships, an extra hundred pounds slammed into an opponent's hull or decking might be enough to turn the tide.

Frigates sailing the high seas usually carried long guns and carronades. Long guns were just that—enormously long and thick-barreled cannon with a narrow bore for long-range action if not terrible accuracy. These varied in caliber from two to forty-two pounds, but were usually of eighteen or twenty-four pounds. Carronades (named for Carron, Scotland, where they were first cast) were much shorter, lighter, and stockier cannon, but were able to throw a much heavier shot for a shorter distance. Carronades usually fired twenty-four, thirty-two, or forty-two-pound shot. Both long guns and carronades were muzzle loaders, meaning that the guns needed to be run in and out on long carriages and their loads rammed down the muzzle with a ramrod between rounds. The load might be a single cannonball of appropriate weight or a canister of grapeshot, an assortment of smaller balls and assorted metal and nails that could wreak havoc with an opponent's crew, rigging, and sails. When both a cannonball and grapeshot were loaded into a gun for close action, it was said to be double-shotted.

The effectiveness of these armaments depended in large

measure on the skill of the ship's captain in maneuvering and keeping to the windward of his opponent. A captain who did so was said to have the weather gauge. Given the rolling decks, cries of "aim for their masts, boys," were more a prayer than anything else, and orders to "mind the weather roll" (the pitching of the deck) were more likely to be heard. All this underscores the importance of broadside weight to the maritime gunnery of the times. If a captain was able to keep the weather gauge and maneuver his vessel so as to deliver a broadside effectively, it was likely to hit something.[4]

When Congress finally declared war on Great Britain, Captain Isaac Hull of the USS *Constitution* was in Washington discussing what role his nation's tiny navy should play in the conflict. There was no easy answer. For starters, Madison's secretary of the navy was Paul Hamilton, a former South Carolina banker and governor whose knowledge of water was largely limited to his experiences as a rice planter. Given Hamilton's lack of seamanship, it is not surprising that everyone had a different idea about how to use the greatly outnumbered American fleet.

Secretary of State James Monroe wanted to keep the handful of ships in port—safe but hardly very effective. With an eye toward the national economy, Secretary of the Treasury Albert Gallatin wanted the ships to stay in American waters and protect returning merchantmen. Hull's compatriots from his Tripoli days, swashbucklers Stephen Decatur and William Bainbridge, were straining at the halyards and begging to be turned loose to cruise separately around the globe. Their theory was that their beefed-up frigates were fast enough to elude any British ship of the line and tough enough to best any frigate or lesser craft they might encounter. Finally, Commodore John Rodgers advocated forming two squadrons of ships and searching out British convoys. Initially President Madison elected to adopt a combination of Gallatin and Rodgers's recommendations: two roving squadrons that would "afford our returning commerce, all possible protection."[5]

This order, however, failed to reach Rodgers before he had gathered a squadron of four ships about USS *President* and sailed from New York in search of British convoys. His force wasn't as outnumbered as it might appear at first blush. To be sure, Great Britain had more than one thousand warships, half of which were at sea at any given time, but as befitted the demands of her empire, those that were not occupied blockading Napoleon's Europe were scattered all over the globe. Consequently, in 1812 the British had only one ship of the line (HMS *Africa*), nine frigates, and twenty-seven smaller vessels operating out of its North American stations at Halifax and Newfoundland. Additional ships were assigned to the West Indies, but they were pretty much tied down there trying to protect Britain's Caribbean trade from all comers.

The British had planned to disperse their ships and blockade American ports, but with Rodgers on the loose, they were forced instead to stick together and search for him. The result was that during the first summer of the war, most American merchant vessels slipped into port safely. New York Governor Daniel Tompkins went so far as to report to his legislature that "Nearly as great a proportion of homeward bound merchantmen have escaped capture as has been customary during the last three or four years of peace."[6]

Not knowing that Rodgers had already sailed from New York, the navy department ordered Hull and the *Constitution* to join him there. The *Constitution* was at Annapolis taking on stores and a new crew, and Hull joined his ship as she sailed down Chesapeake Bay. On July 12, 1812, with a crew of about 450, *Constitution* passed through the Virginia Capes and turned northward up the eastern coast.

Meanwhile, the thirty-six-gun frigate HMS *Belvidera* en route from New York to Halifax with news of the declaration of war had barely managed to elude Rodgers in the *President*. Upon *Belvidera*'s arrival in Halifax, British Vice Admiral Herbert Sawyer consolidated his forces and dispatched a powerful squadron under

Captain Philip Broke to pounce on Rodgers. Broke flew his flag from the thirty-eight-gun frigate HMS *Shannon*. With *Shannon* sailed the sixty-four-gun ship of the line HMS *Africa*, the thirty-two-gun frigate HMS *Aeolus*, and the *Belvidera*. Four days later while off Nantucket, they were joined by HMS *Guerrière*, a thirty-eight-gun frigate under the command of Captain James Richard Dacres. This was the ship that Rodgers and the *President* had sought before firing in error upon the *Little Belt* some months before. Dacres was determined not to let the Americans forget the incident, and upon *Guerrière*'s foresail and mainsail were emblazoned the words, "*Guerrière*, not the *Little Belt*."[7]

Inbound to New York, Hull sighted four ships sailing westward and a fifth heading straight for the *Constitution*. Hull judged that the vessels might be those of Commodore Rodgers and his squadron, but he was cautious. The breeze was very light and as night fell, Hull hoisted prearranged signal lights. When the ship bearing down on him failed to answer and identify herself, Hull ordered the *Constitution* to turn away and wait for daylight to assess the situation. But daylight did not bring good news. The detached vessel proved to be the arriving *Guerrière* and the other four vessels the remainder of Broke's squadron. Where Rodgers was, was anyone's guess, but now Hull was in deep trouble. The sea around the *Constitution* was calm, while the British ships came on with a good following breeze.

Hull would have been in even more trouble had not Dacres also become worried about signals during the night and held *Guerrière* off on another tack, fearful that Broke's four ships might actually be Rodgers's squadron rather than his own planned rendezvous. When Dacres finally realized his error, he had lost his chance to trap *Constitution* between the *Guerrière* and Broke's ships, and his contempt for the American navy shot up another couple of notches.

By eight o'clock the next morning, *Shannon*, *Africa*, *Aeolus*, and *Belvidera* were almost within firing range of the *Constitution*. But now the breeze died for both navies. Hull put his long boats in the water and ordered their crews to bend their backs and row for all they were worth in an attempt to tow the *Constitution* forward. The British ships quickly did likewise. Determined to fight it out at any cost, Hull next ordered two twenty-four-pounders moved from the gun deck and run out the rear windows of his cabin to point astern. He also lightened his ship by pumping almost ten tons of drinking water into the sea. Equally determined that this Yankee upstart would not get away, Captain Broke soon ordered all the long boats in his squadron to tow the *Shannon*, and she soon began to close the gap.

Then Hull's executive officer, Lieutenant Charles Morris, had another idea. The *Constitution* was in 22 fathoms of water (about 130 feet). Why not try kedging, the practice of rowing an anchor ahead of the ship, dropping it to the bottom, and then winching the ship forward by taking up the slack. Hull had seen the maneuver plenty of times on the Housatonic River of his childhood, and he readily agreed. Meanwhile, *Guerrière* was inching closer from the opposite side. She fired a broadside, but it fell well short. Finally, by the third day of this frenzied race of tortoises, a light breeze began that benefited *Constitution* first. Slowly but steadily her canvas filled and she began to pull away from the British ships. It took another day, but one by one the topsails of the British frigates gradually dropped below the horizon.

What a chase and what a close call for the Americans! Now Hull made not for New York but for Boston, where rumors of the *Constitution*'s demise unrelated to her escape from Broke were rampant. The ship's appearance in Boston Harbor and the true story of her escape from four British frigates and a man-of-war filled the streets and pubs. Hull was the toast of the town, but his stay in port was short-lived. Determined to avoid being blockaded, Hull replenished his supplies and armaments and quickly put to sea again on August 2. Soon he would be the toast of the nation. Meanwhile, Captain Broke summoned his commanders aboard *Shannon* to account for the failure of the chase. Captain Dacres complained bitterly that the other ships had not properly answered his signal lights that first night and then vowed that he would get revenge upon the Americans.[8]

Leaving Boston harbor, Hull steered the *Constitution* northeast along the coasts of Nova Scotia and Newfoundland and took up station off Cape Race in the Gulf of St. Lawrence—the route of Canada's lifeline to Great Britain. *Constitution* captured and sank two British merchantmen there and then turned south toward Bermuda. She soon chased down a brig that turned out to be the American privateer *Decatur*, of fourteen guns. Her captain reported to Hull that he had outrun a British frigate the day before. This was good news to Hull. Now he knew that there was a British warship in the vicinity and that he had just run down a vessel that had proven faster the day before.

About two o'clock on the afternoon of August 19, 1812, the *Constitution*'s lookouts raised cries of "Sail ho" and pointed to a sail bearing east-southeast. Pulses quickened. Having the wind, the *Constitution* gave chase and quickly closed to within three miles. Both ships beat to general quarters, but Hull was still uncertain about the identity of his quarry. It soon proved to be the frigate *Guerrière*, and her captain was about to get his wish for a chance at revenge. Supposedly Captain Dacres was quoted as boasting, "There is the Yankee frigate: in forty-five minutes she is certainly ours: take her in fifteen and I promise you four months' pay."[9]

As *Constitution* closed with *Guerrière* from its windward side, an overeager Dacres ordered his crew to fire broadside after broadside at the approaching vessel. By and large, Hull held his fire, much to the consternation of Lieutenant Morris, who asked three times for permission to do so. British cannonballs struck the *Constitution*, but did little damage. After one particularly well-aimed British broadside bounced harmlessly off *Constitution*'s hull, a crew member is reported to have exclaimed, "Hurrah, her sides are made of iron," or words to that effect. No matter. A legend was born, and "Old Ironsides" she became.

Finally, with the two ships but twenty-five yards apart, Hull

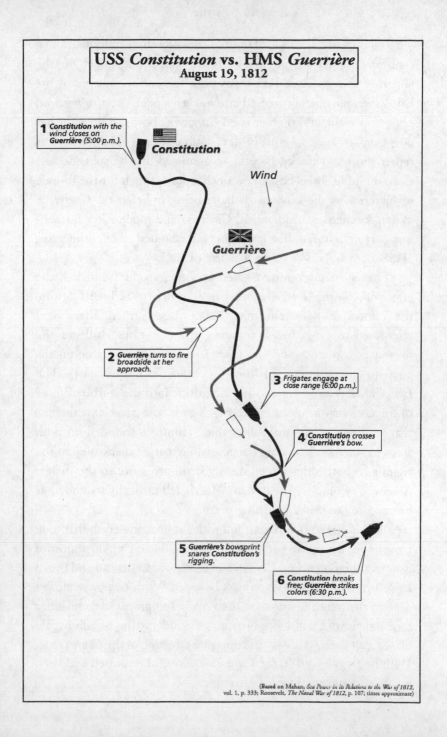

USS *Constitution* vs. HMS *Guerrière*
August 19, 1812

1 *Constitution* with the wind closes on *Guerrière* (5:00 p.m.).

Constitution

Wind

Guerrière

2 *Guerrière* turns to fire broadside at her approach.

3 Frigates engage at close range (6:00 p.m.).

4 *Constitution* crosses *Guerrière*'s bow.

5 *Guerrière*'s bowsprint snares *Constitution*'s rigging.

6 *Constitution* breaks free; *Guerrière* strikes colors (6:30 p.m.).

(Based on Mahan, *Sea Power in its Relations to the War of 1812*, vol. 1, p. 333; Roosevelt, *The Naval War of 1812*, p. 107; times approximate) ordered his first broadside of double-shot—both a cannonball and a canister of grape—from his starboard guns. The effect on the *Guerrière* at such close range was dreadful, and the cheers of the British seamen quickly quieted into moans of pain. *Constitution* fired again and again, and then crossed *Guerrière*'s bow and brought her port guns to bear in equally devastating fashion. Hull, who was rather short and stocky, became so animated that he split the seat clean out of his breeches, an event that did as much for the morale of his crew as the obvious damage being inflicted on *Guerrière*. When another broadside raked *Guerrière* and toppled her mizzen-mast, Hull ignored his breeches and shouted above the roar, "Huzza, my boys! We've made a brig of her."

The fallen mizzenmast acted as a huge rudder and had the effect of slowing *Guerrière* and swinging her to starboard despite the efforts of her helmsman. *Constitution* surged ahead and attempted to cross her bow again, but this time Hull cut the maneuver too close and the *Guerrière*'s bowsprit became entangled in the rigging of the *Constitution*'s mizzenmast. The *Constitution* poured yet another broadside into the starboard bow of the *Guerrière* while the *Guerrière*'s own bow guns landed shots that set fire to Captain Hull's cabin. Trumpets sounded on both ships to summon boarding parties, while marine marksmen in the rigging of both ships added deadly small-arms fire to the melee. Aboard *Constitution*, Lieutenant Morris fell critically wounded as he prepared to lead a boarding party.

Then *Guerrière*'s foremast fell with a splintering crash that took most of her mainmast with it. The ship shuddered and lost most of its forward momentum. *Constitution* continued under sail and broke loose from the grip of *Guerrière*'s bowsprit. With *Guerrière* almost dead in the water, *Constitution* drew apart and prepared to rake her fore and aft with still more broadsides. Suddenly the British frigate fired a shot to leeward—in the opposite direction of the *Constitution*. With no flags left to strike, Captain Dacres, who himself had been wounded, was signaling his surrender.[10]

Captain Hull sent Lieutenant George C. Read aboard *Guerrière* to ascertain the situation. Stepping across decks slippery with blood, young Read confronted Dacres amid the carnage of his quarterdeck and inquired, "Captain Hull presents his compliments, sir, and wishes to know if you have struck your flag?" To this, Dacres is said to have replied, "Well, I don't know. Our mizzen mast is gone, our fore and main masts are gone—I think on the whole you might say we have struck our flag."

Dacres was escorted aboard *Constitution* to confront Hull, who refused Dacres's tender of his sword, supposedly saying that he could not take the sword of one who had defended his ship so gallantly. Then Hull asked if there was anything on the *Guerrière* that Dacres wished to have brought aboard. "Yes," Dacres replied, "my mother's Bible." Hull ordered it retrieved.[11]

Constitution had sustained losses of seven killed and seven wounded to British losses of fifteen killed and seventy-eight wounded—the latter number due in no small measure to Captain Hull's use of grapeshot at close range. Hull hoped to tow the *Guerrière* into port as a prize, but by dawn the next morning there was four feet of water in her hold. By midafternoon Hull recalled his prize crew and ordered her blown up. In transferring the crew of *Guerrière* to the *Constitution*, Hull found that there were ten impressed Americans aboard—a graphic example of one of the war's causes. Dacres clearly knew of it because he had graciously permitted the Americans to go belowdecks rather than fight against their countrymen.[12]

Hull made sail for Boston to report his victory and land his prisoners. By itself, the loss of one British frigate was not important, but it was to have immense psychological impacts on both sides of the Atlantic. Details of Captain Hull's victory were printed in Boston newspapers on September 2, 1812—near columns that told of his uncle's retreat from Canada.[13] Might Federalist New England have thrown up its hands and seceded after General Hull's subsequent defeat at Detroit, if not for his nephew's victory? Perhaps. But instead of secessionist talk, New England suddenly rallied around a grand naval victory on the very seas it depended on. New England—and indeed the entire nation—went wild with adoration for Captain Hull, his crew, and the ship that everyone soon called "Old Ironsides." Gone, too, was the disinterest Congress had shown in the American navy. Even the war hawks, who had managed only a measly $300,000 to repair three frigates just the preceding spring, now voted $2.5 million to build four seventy-four-gun ships of the line and six forty-four-gun frigates.

And Hull's victory had political ramifications as well. There is nothing like a victory to galvanize voters to a wartime incumbent president. Studious James Madison reaped the political benefits of the *Constitution*'s broadsides and rode hesitantly but decisively to a second term the following November. And the loss of Detroit? Hey, Henry Clay and his western cohorts were voting for Madison regardless; it was the swing votes of the eastern states that the *Constitution*'s victory pushed into Madison's column.[14]

Meanwhile, across the Atlantic, conveniently forgetting John Paul Jones's capture of HMS *Serapis* during the Revolution, the *London Times* bemoaned that "Never before, in the history of the world, did an English frigate strike to an American." While hesitant to condemn Dacres outright, the newspaper nonetheless went on to say that "there are commanders in the British Navy who would a thousand times rather have gone down with their colours flying, than to have set their brother sailors so fatal an example."[15]

Dacres faced the inevitable court of inquiry and came up with a novel defense. The *Guerrière* was originally built as a French frigate and was captured by the British in 1806. Clearly, Dacres maintained, she wasn't as sturdy as those frigates built in British shipyards. He claimed that the *Guerrière*'s masts were rotten and that they had toppled from this defective condition rather than from the direct fire of the *Constitution*. Then, too, he noted that he had been handicapped by excusing the impressed Americans—

less than 5 percent of his crew—from action. Incredibly, he was exonerated, but hardly humbled. Apparently forgetting the humiliating moment before Lieutenant Read on *Guerrière*'s quarterdeck, Dacres could not resist a final broadside. "I am so well aware that the success of my opponent was owing to fortune," he declared, "that it is my earnest wish to be once more opposed to the *Constitution*, with the same officers and crew under my command, in a frigate of similar force to the *Guerrière*."[16] That was one favor the Royal Navy was not about to accord him.

Isaac Hull's adoring biographer, Bruce Grant, asserted that "the victory of Old Ironsides over the *Guerrière* and the importance of the man who commanded this American frigate cannot be measured in terms of the ordinary sea battle." He was right. It was a huge shot of adrenaline into a national consciousness that was still groping for purpose. "In the midst of battle," Grant continued, "the nation's lack of faith in the Navy as a fighting arm vanished; and when the *Guerrière* fired her gun to leeward in submission it was the signal for a new naval era in America."[17]

In retrospect, history generally supports that view, but in the fall of 1812—no matter how glorious the *Constitution*'s triumph—it was still hard to argue with the final assessment of the *London Times*. "The war had lasted nearly four months;" the paper observed, "yet, with the exception of the fatal, and ever-to-be-lamented surrender of the *Guerrière*, the record of the American achievements is a universal blank."[18]

Universal blank or not, America's record was about to get another hero. Stephen Decatur was born in Sinepuxent, Maryland, on January 5, 1779, and made his first ocean voyage as a cabin boy en route to France before he was ten. By 1798 Decatur was a midshipman aboard the *Constitution*'s sister ship, the forty-four-gun frigate USS *United States*. He later served under Captain William Bainbridge on the USS *Essex* and then commanded the schooner USS *Enterprise* off the coast of Tripoli. There he made his reputation as somewhat of a daredevil by sail-

ing the captured ketch *Intrepid* into the harbor at Tripoli and burning the USS *Philadelphia* to prevent its use by Barbary pirates after
the ship ran aground and was captured. Long before he became
the scapegoat of the *Chesapeake* affair, James Barron appointed the
up-and-coming Decatur to command the *Constitution* for a short
time. Like Isaac Hull, Decatur, too, would later sit on Barron's
court of inquiry.

On October 8, 1812, Commodore John Rodgers left Boston
on a second cruise in search of British convoys, hoping that it
would be more fruitful than his empty-handed crisscrossing of the
Atlantic three months before. With his flagship USS *President*
sailed the thirty-eight-gun frigate USS *Congress*, the sixteen-gun
brig USS *Argus*, and the veteran *United States*, now under
Decatur's command and packed with ten more cannon than her
standard forty-four-gun rating. The *Argus* and the *United States*
soon took leave of Rodgers and went their separate ways.

The *United States* sailed eastward across the Atlantic and on
October 25, some six hundred miles south of the Azores and a like
distance west of the Canaries, chanced upon a single sail. It
proved to be the thirty-eight-gun frigate HMS *Macedonian* bound
for station in the West Indies and sporting a beefed-up complement of fifty guns. Unlike the *Guerrière*, the *Macedonian* was not a
rebuilt French prize, but was newly constructed out of stout
British oak. If things went poorly, her captain, John Surnam
Carden, would have to look for other excuses.

In just the reverse of the *Guerrière–Constitution* situation, the
British frigate held the initial advantage of being to windward.
Captain Carden seemed unable to make the most of it, however,
while brash Decatur boldly ignored it. The *Macedonian* was reputed
to be a crack ship with a crack crew that Carden had drilled incessantly in gunnery. Yet as Carden appeared preoccupied with keeping
the wind while closing with the *United States*, it was Decatur who
repeatedly eased his ship off the wind and used the occasions to rake
the oncoming British frigate with long-range broadsides.

Decatur continued to play this game of catch-me-if-you-can while his gunners fired with surprising accuracy. The *Macedonian* never got within less than one hundred yards of the American frigate, and there was little exchange of grapeshot and musketry. In fact, as Decatur dueled at long range, the U.S. marines aboard the *United States* had little to do. Finally *Macedonian*'s mizzenmast was shot away at the deck and the tops of her fore- and mainmasts fell as well.

The *United States* came about behind the *Macedonian*, fired more broadsides into her stern, and then crept up on her leeward side. Carden had kept his wind, but he also now had more than one hundred cannonballs in his ship's hull. Prematurely, some would later say, Carden struck his flag. His crew suffered 43 killed and 71 wounded out of a complement of 301. The *United States* sustained casualties of only 7 killed and 5 wounded out of a crew of 428. Broadside weight and the skill to deliver it had done the damage. According to Theodore Roosevelt, the *United States* was throwing 786 pounds to the *Macedonian*'s 547 pounds. As British seaman Samuel Leech later recalled, "Grapeshot and canister were pouring through our portholes like leaden hail; the large shot came against the ship's side, shaking her to the very keel. . . ."[19]

After her surrender the *Macedonian* was in far better shape than the *Guerrière* had been. Decatur promptly dispatched a prize crew and triumphantly sailed the British frigate back across the Atlantic and into Newport harbor, the only British frigate ever brought into an American harbor as a prize of war. Carden was in no hurry to get there because he knew that once exchanged, the inevitable court of inquiry awaited him. He also knew the Royal Navy's record. In twenty years of almost incessant naval warfare and numerous single engagements between French and British frigates, only once had the French been victorious. That had been in 1807 when the much heavier *Milan* bested HMS *Cleopatra*. The British were not only unaccustomed to losing, they were scarcely aware of the possibility. Now on the heels of the

Guerrière disaster came news that Stephen Decatur was bringing a British frigate into port as a prize.[20] What in the world, the British Admiralty wondered, was going on?

While Stephen Decatur and the crew of the *United States* took their bows as America's latest heroes, the *Constitution* was again at sea. After dispatching *Guerrière*, Isaac Hull—toasted hero that he was—asked for a short leave to deal with the death of his brother. Thirty-eight-year-old William Bainbridge was given command of the *Constitution* in his absence. Bainbridge's naval career thus far had had its less than glorious moments. As a young lieutenant in command of the schooner USS *Retaliation*, Bainbridge had once surrendered to two French frigates. As captain of the USS *Philadelphia*, he had been the one to run her aground under the guns of Tripoli. But in some respects, such fates were deemed part of the growing pains of the young American navy. Thus, with Hull on leave, Bainbridge was given command of a small squadron that was to include his old ship, USS *Essex*, and James Lawrence's sloop, the eighteen-gun USS *Hornet*. When only the *Hornet* was ready to sail, *Hornet* and *Constitution* left Boston on October 26, 1812, bound for the South Atlantic.[21]

After taking on fresh water in the Cape Verde Islands, both *Hornet* and *Constitution* appeared off the Brazilian port of Bahia—now called Salvador. In the harbor was HMS *Bonne Citoyenne*, an eighteen-gun sloop under the command of Captain Pitt Barnaby Greene. She was a perfect match for the *Hornet*. Captain Lawrence promptly sent a challenge into Bahia, proposing to Captain Greene that his ship come out and fight the *Hornet*. In the convention of the day, Lawrence gave his word that the duel would be between only the two smaller ships. The *Constitution*, Lawrence assured Greene, would not take part in the affair.

What Greene thought of the challenge is perhaps best summed up by the fact that the *Bonne Citoyenne* remained safely anchored in Bahia's neutral harbor. Perhaps Greene doubted Lawrence's assurance that the *Constitution* would not intervene.

More likely, he chose to safeguard the half million pounds of British sterling that were in his hold. Whatever the reason, Greene's refusal to fight was—by Theodore Roosevelt's count—the only time during the war that a ship of relatively equal force of either navy declined to accept such a challenge. So with Greene determined to sit it out in the safety of Bahia, Bainbridge ordered Lawrence and the *Hornet* to blockade *Bonne Citoyenne*, while he sailed the *Constitution* in search of bigger fish.

It didn't take long to find one. On the morning of December 29, 1812, *Constitution* was about thirty miles off the Brazilian coast when two sails were spotted bearing down on her with the wind. One proved to be the captured American merchantman *William*, but the other was the thirty-eight-gun frigate HMS *Java*, outbound from England and en route to a station in the East Indies. Aboard was the newly appointed governor of India, Lieutenant General Thomas Hislop, and his retinue. This fact did not, however, stop *Java*'s captain, Henry Lambert, from moving to engage the American frigate.

Like HMS *Guerrière*, the *Java* was a captured French frigate. Originally named *Renommée* and taken by the British off Madagascar in May 1811, the ship had been refitted and then renamed to celebrate the capture of that East Indian island from the Dutch. Captain Lambert had heard of *Guerrière*'s defeat before leaving England, and perhaps he was eager to prove her fate a fluke. *Java* raced before the wind toward *Constitution* in much the same manner Hull had employed against *Guerrière*. Unlike Dacres, Bainbridge was content to stand off some and take time to size up his opponent.

By early afternoon, both ships had closed and shortened their sails in preparation for battle. *Constitution* opened with a broadside that did little damage. *Java*'s first broadside was much more destructive. Lambert kept edging his ship closer, taking advantage of her faster speed in an attempt to cross *Constitution*'s bow and rake her. But Bainbridge was equal to the challenge. Against convention, he reset his main- and foresails—usually considered too unwieldy for such close action—and managed to keep abreast of the British frigate, preventing her from crossing his bow while engaging in furious broadsides. Such action was not taken with impunity. A British cannonball shot away the *Constitution*'s wheel, wounding Captain Bainbridge in the process.

Now Captain Lambert put *Java*'s helm over to bring her alongside the *Constitution* in preparation for boarding. Musketry and small-arms fire added to the fight. But the heavier American broadside weight—654 pounds to 576 pounds—was beginning to take its toll. By the time the Americans poured grapeshot into the fray, *Java*'s rigging and masts were in disarray. Her mizzenmast fell, the stump of her crumpled foremast was blown away, and soon her mainmast came crashing down as well. One of the marines in *Constitution*'s rigging fired a shot that wounded Captain Lampert. He would not live to face his own court of inquiry.

With *Java*'s masts all down, Bainbridge ordered *Constitution* to stand off slightly and repair her own damage. *Java* was not going anywhere. Someone hoisted the Union Jack on what little was left of one of *Java*'s masts, but it was only for show. When *Constitution* came back alongside her, Lambert struck his colors and Bainbridge dispatched a prize crew to board her. Bombay-bound Governor Hislop presented Bainbridge with a sword and thanked him for his courtesies. All in all, the fight had been more severe than that with the *Guerrière*, but once again there was no question but that "Old Ironsides" had been victorious.[22]

Not knowing of Decatur's triumph in taking the *Macedonian* into port as a prize, Bainbridge attempted to do the same with *Java*, but the British frigate was soon wallowing like a badly wounded whale. On New Year's Eve, 1812, Bainbridge ordered her crew removed and the hulk set afire. As she slid beneath the warm waters of the South Atlantic, the Americans cheered while the incredulous British added another name to a growing list of maritime defeats. Rather than the indignation that had met the sinking of the *Guerrière*, the subject of the *Java*'s loss had become, in the editorial comment of the British *Naval Chronicle*, "too painful for us to dwell upon."[23]

Thus, as 1812 drew to a close, the "universal blank" of the American record had two more significant victories upon it. Britannia still ruled the waves, but as incompetent as the American army had proven, the American navy, with a handful of frigates ably captained by officers who knew how to fight, had given the country something to cheer—even in flinty New England. On their face, these naval victories would be of little strategic importance to the outcome of the war, but memories of the bitter defeats would heighten British resolve—and retribution—when the lion finally turned its full attention to North America. Meanwhile, the glories of the *Constitution* and the *United States* lifted American spirits and quickly became the stuff of legend.

Marching on a Capital

The spring of 1813 found America's thirst for Canada unabated. Any hopes for its easy conquest, however, had been sorely tempered by the failed campaigns of the previous summer. Even now, William Henry Harrison's march to recapture Detroit had been blunted at the River Raisin and New England remained as reluctant as ever to support a major drive on Quebec. That left Niagara. So, as spring brought some measure of renewed optimism, the three-prong American attack of 1812 was replaced by one concerted effort against Lake Ontario and the Niagara frontier in the expectation that at the very least Upper Canada might be wrested away from the British crown. Exactly how and where this thrust was to come was a matter of some debate.

In the big scheme of things, disgraced General William Hull had been right. The control of the Great Lakes was essential to any effective military operations along the entire Northwest frontier. Within weeks of Hull's surrender at Detroit, Secretary of the Navy Paul Hamilton ordered Isaac Chauncey, a forty-year-old veteran navy captain, "to assume command of the naval force on Lakes Erie and Ontario, and to use every exertion to obtain control of them this fall."[1]

Born in Connecticut in 1772, Chauncey seemed bound for a career in law when the sea lured him away from home at the age of twelve. He never looked back, and by nineteen was captaining ships for fur-trade mogul John Jacob Astor. After the American navy was formed, Chauncey received a commission as a lieutenant in 1798 and supervised the construction of USS *President* in New York harbor. Like many officers of his generation, Chauncey saw service off Tripoli, first as the executive officer on the *President* and later on USS *Chesapeake*.

After the Tripoli campaigns, Chauncey went back to work for Astor. In 1806, he was in command of Astor's merchant ship *Beaver* when the 64-gun HMS *Lion* pulled alongside and demanded that Chauncey muster his crew—the usual prelude to a press gang. When Chauncey refused, the British captain threatened to impress his entire complement. Do it, Chauncey shrugged. Suddenly aware of creating an international incident well beyond normal impressment antagonisms, the British officer backed down and the *Beaver* continued on its way with a full crew. Upon his return home in 1807, the navy offered Chauncey command of the New York Navy Yard, a post he held until tapped for service on the Great Lakes.[2]

Chauncey's charge to obtain control of Lakes Erie and Ontario would require his considerable shipbuilding skills as much as, or more so than, his combat experience. The only American ship of consequence on Lake Ontario at the time was the eighteen-gun brig *Oneida*. The only armed American vessel on Lake Erie had been the fourteen-gun brig *Adams*, but it was captured at the fall of Detroit. Renamed *Detroit*, it now flew the British ensign. Two lakes, one ship. Chauncey had his work cut out for him.

Either Secretary Hamilton expected miracles or he had no concept of the difficulties involved in assembling a workforce of shipwrights on the lakes, let alone launching two fleets—one for Lake Ontario below Niagara Falls and one for Lake Erie above them. "You will have a number of vessels to build, and the timber is yet to be cut," Hamilton advised Chauncey. "We hope, that in twenty days from the time of your arrival at Sackett's Harbour & at Buffaloe [*sic*], your vessels will be ready to act. You will have to build of green timber; but it is a case of necessity, and cannot be avoided. . . . At all event, be the cost what it may, we must have possession of the Lakes this fall."[3]

That didn't happen, of course, but Captain Chauncey did make significant shipbuilding progress by the end of 1812. At Sackets Harbor, New York, on Lake Ontario, he succeeded in launching the twenty-four-gun USS *Madison* (a three-masted *ship* almost twice the size of the *brig Oneida*) and converting a number of private lake schooners into gunboats mounting one or two guns.[4] Lake Erie was a different matter.

To run the show on Lake Erie, Chauncey chose Jesse Duncan Elliott, a thirty-year-old lieutenant who had joined the navy in 1804. There had been nothing remarkable about Elliott's career, save for the fact that as a young midshipman he had been on the deck of the *Chesapeake* when HMS *Leopard* opened fire. Chauncey sent Elliott to Buffalo to find "the best Site to build, repair, and fit for service such Vessels or Boats as may be required to obtain command of Lake Erie."[5]

New York militia commander General Stephen Van Rensselaer, who was then still in command at Buffalo, was lukewarm to the whole idea of a Lake Erie fleet. He thought that the navy's efforts should be directed solely against Lake Ontario. New York Congressman Peter B. Porter came to Elliott's rescue and suggested his family's shipyard at Black Rock on the Buffalo side of the Niagara River just downstream from Lake Erie. Elliott set about converting several schooners there, but it quickly became apparent that Black Rock had two major drawbacks. First, any ships that sailed upstream from Black Rock into Lake Erie had to fight several miles of the river's four-mile-per-hour current. Second, they had to do so under the guns of Fort Erie on the Canadian side. General Van Rensselaer promised to capture Fort Erie, but that promise fell flat after the Americans were repulsed at Queenston Heights.[6]

While pondering his options, Elliott watched the British brigs *Detroit* (formerly the American *Adams*) and *Caledonia* (the North West Company vessel that had already lent a hand in the capture of Mackinac Island) anchor below Fort Erie. Figuring that it was a lot easier to capture a vessel in these waters than it was to build one, Elliott boldly assembled about fifty sailors and a like number of army regulars, and in the dark of night floated down on the vessels and took their crews by surprise. For a moment Elliott pondered sailing the two brigs into Lake Erie and taking on a third British vessel, but the current of the Niagara River was too strong in the light night winds, and both ships drifted downriver toward Black Rock. In the process the *Detroit* ran aground, and amid a bombardment from Fort Erie, Elliott ordered her burned lest she be recaptured. The *Caledonia* made it safety to Black Rock. That was one ship that Chauncey wouldn't have to build.[7]

Meanwhile, unbeknownst to either Chauncey or Elliott, a Great Lakes merchant captain named Daniel Dobbins—who had set something of a record by managing to be captured by the British at the fall of both Mackinac and Detroit—was making the rounds in Washington. In an example of how terribly loose and disorganized the nation's chain of command was, Dobbins convinced Secretary Hamilton that with Black Rock coming under fire, the perfect place to build ships on Lake Erie was at Presque Isle Bay near his home at the tiny village of Erie, Pennsylvania. Hamilton approved Dobbins's plan without consulting Chauncey. Elliott received the order directly from Hamilton and went through the roof, strongly arguing that there was not sufficient water over the bar at the mouth of Presque Isle Bay to allow any large ships built there to cross into the lake. Nonsense, replied Dobbins, who had hastened to Erie, "I believe I have as perfect a knowledge of the lake as any other person on it and I believe you would agree with me if you were here."[8]

Perhaps because he expected sparks to fly with Dobbins, Chauncey recalled Elliott to Lake Ontario and gave him command of the *Madison*. To deal with Dobbins and build a navy from scratch at Presque Isle, Chauncey chose another young lieutenant who had been champing at the bit for some action while commanding gunboats at Newport, Rhode Island. His name was Oliver Hazard Perry, and in time he would also come to question Dobbins's self-proclaimed knowledge of the lake.

• • •

By now President Madison, his second election secure, had shuffled his cabinet. Secretary of War William Eustis became the scapegoat for the failed land campaigns of 1812 and resigned. In his place, Madison appointed John Armstrong of New York. Armstrong had served in the Revolution and was currently a brigadier general in command of New York City. Armstrong was not Madison's first choice, and after he was narrowly confirmed by a Senate vote of 18 to 15, it was clear that he was hardly anyone's first choice. Secretary of State James Monroe was among those who were less than pleased with Armstrong. Monroe was certain that Armstrong had political ambitions of his own that might disrupt the tradition of Virginian succession to the presidency that had Monroe next in line.

Secretary of the Navy Paul Hamilton was also out. His replacement, former Pennsylvania congressman William Jones, was better received and was in fact Madison's first choice. Unlike Hamilton, Jones actually had extensive maritime experience. Jones determined to give Chauncey a free hand on the Great Lakes and told him that "you are to consider the absolute superiority on all the Lakes, as the only limit to your authority."[9]

That wasn't quite true, of course. For one thing, Chauncey's dominion only extended to his still fledgling navy. That meant that whatever he did had to be coordinated with the army. Fortunately, newly appointed Secretary of War Armstrong was enough of a soldier to understand the strategic importance of the St. Lawrence River. One thrust aimed squarely at Lake Ontario, Armstrong believed, would chop off the limbs of Upper Canada from their St. Lawrence trunk and open up an avenue down which to attack Montreal.

In preparation for such an assault, Armstrong ordered General Henry Dearborn—who had dawdled so in New England the summer before and who was "Granny" to his troops more than ever—to position a brigade of three thousand men at Buffalo and a second of four thousand men at Sackets Harbor. The Buffalo contingent was to attack Fort George and Fort Erie to gain control of the

Niagara inlet to the lake, while the Sackets Harbor force was to seize Kingston, Ontario, at the lake's St. Lawrence outlet.[10]

It was actually a pretty good plan, but like almost every American strategy on land during the war, its execution in the field left much to be desired. Scarcely had Dearborn and Chauncey put their heads together than they began to have second thoughts about the attack on Kingston. Chauncey may have been concerned that the spring buildup of ice at the lake's outlet to the St. Lawrence would interfere with his ships coming into Kingston harbor. Dearborn should have been delighted that his objective was barely thirty miles from Sackets Harbor, but instead "Granny" fretted over the numbers of the opposing troops. (In truth, the British were sitting in Kingston doing exactly the same fretting, and the forces were relatively matched.)

No, Kingston was too risky, Chauncey and Dearborn decided. Why not make a raid on York instead? York, Ontario—later to become Toronto—sat on the northwestern shores of Lake Ontario and was of dubious strategic value in and of itself. York was, however, the capital of Upper Canada, and Chauncey had received reports that in its shipyard the British were constructing two new brigs of eighteen guns each, one of which was to be named for the hero of Queenston Heights, Sir Isaac Brock. As Lieutenant Elliott had proven with the *Caledonia*, it was easier to capture ships than to build them.

But was this watching the hole instead of the donut? Kingston was the cork in the bottle of the Great Lakes, the gateway into and out of the St. Lawrence. York was a poorly defended harbor that would prove as indefensible for the attacking Americans as it was for the British, and it was the gateway to nowhere. Fort George and Kingston each controlled ingress and egress to the lake. York controlled nothing. It was certainly not on the trunk of the St. Lawrence–Great Lakes artery, but merely one of its minor upper branches. What good was York if the heart of Upper Canada around Lake Ontario still stood securely anchored to the St. Lawrence trunk? With that point apparently clear only in

Secretary Armstrong's mind, the American army and navy nonetheless sailed westward from Sackets Harbor to attack York, some 160 miles away.[11]

Aboard Chauncey's little navy as it departed Sackets Harbor on April 25, 1813, were about fifteen hundred regulars under the direct command of a newly appointed brigadier general named Zebulon Pike. Born in Trenton, New Jersey, in 1779, this career officer was a long way from the geography that history normally associates with him. In the wake of the Louisiana Purchase, Pike's first assignment as a young lieutenant had been to follow the Mississippi River to its source. He came close, arriving at north-central Minnesota's Leech Lake instead of Lake Itasca. Then in July 1806, Pike and twenty-two men headed west from St. Louis. On their face, his orders were straightforward: follow the Arkansas River to the Rocky Mountains, locate the source of the Red River, and follow it back east, roughly defining the southern boundary of the new territory. Nothing, however, was ever straightforward about the man who gave him those orders—none other than General James Wilkinson.

Aaron Burr's plot to steal an empire was in high gear that year, and some have assumed that Pike was a pawn in Wilkinson's machinations with Burr. Whispers of Pike's involvement with Wilkinson's various schemes were only fueled by the fact that Pike seems to have been one of the few regular army officers to get along with Wilkinson and hold him in some measure of esteem.

It didn't help suspicions that Pike and his men made a circuitous loop through the Colorado Rockies, failed to find the Red River, and were apprehended by Spanish officials as trespassers if not outright spies well beyond the boundary of the Louisiana Territory. When Pike finally saw the Red River, it was upon the Spanish returning him to American territory at Natchitoches, Louisiana, in July 1807. His journals and maps were confiscated and not returned to American archives for more than a century. Circuitous wanderings, capture by a foreign power, loss of his records, and a dubious association with the darkest American character since Benedict Arnold—no wonder that Zebulon Pike has frequently been given short shrift beside Lewis and Clark!

But whatever glory as an explorer would later be bestowed on him, Zebulon Pike was still looking for glory as a soldier as his troops rode the icy waves of Lake Ontario toward York. The night before sailing, Pike wrote to his father, who had led troops during the Revolution, "I embark to-morrow in the fleet, at Sackett's Harbor, at the head of a column of 1500 choice troops, on a secret expedition. Should I be the happy mortal destined to turn the scale of war, will you not rejoice, oh my father? May heaven be propitious, and smile on the cause of my country. But if we are destined to fall, may my fall be like Wolfe's [at Quebec in 1759]—to sleep in the arms of victory."[12] To his troops, Pike was less sublime. He issued strict orders that there was to be absolutely no plundering of private property. Violations would result in death, because, Pike explained, "the unoffending citizens of Canada are many of them our own countrymen, and the poor Canadians have been forced into this war."[13]

On the morning of April 27, Pike watched from the deck of Chauncey's flagship, *Madison*, as the first wave of his troops went ashore about three miles west of York. The American invaders outnumbered the British and Canadian defenders at least two to one, but a British-led company of Indian allies initially poured a withering fire onto the beachhead. The prompt arrival of a company of Glengarry Light Infantry might have pushed this first American wave back into the lake, but somehow it inexcusably took a wrong turn en route to the beachhead and arrived too late. By then, Pike was ashore with a second wave of three companies of infantry and had them formed up and moving eastward.

British resolve quickly melted away, and as it did so, Major General Roger Hale Sheaffe, who had come to Brock's rescue at Queenston Heights, withdrew his regular troops out of the town and retreated toward Kingston, leaving the local militia to offer token resistance. Later, some would remember that Sir Roger had been born of American parents in Boston and question the haste of his departure. Given the poor state of York's defenses, however, he probably had little choice.

So Pike's troops—with the aid of broadsides from Chauncey's fleet—captured two aging artillery batteries and advanced to the base of the ramparts of York's western garrison. As his men formed to make the final assault, Pike sat down on a stump to question a captured Canadian militia sergeant. "What is your whole strength and where are the regulars?" Pike demanded. "Sir," the prisoner replied indignantly, "I am a British soldier."[14]

In that instant, a terrible roar came from the ramparts above. The retreating Canadians had detonated the underground powder magazine that held barrels upon barrels of gunpowder and shot. The result was an explosion of volcanic proportions that shook the earth and sent a flurry of huge rocks and shells hurtling about like gigantic grapeshot. Pike's prisoner was killed instantly. Pike turned slightly to shield himself from the blast, but was struck in the head and back by a huge boulder. There was no question that the injury was mortal. Pike the explorer who had tried to climb the Colorado peak that now bears his name was killed by rockfall on the shores of Lake Ontario. With him seems to have died any semblance of American discipline.

"Granny" Dearborn, who had been aboard the *Madison* in overall command, now lumbered ashore to assess the situation. With Pike dead and American casualties running high from the magazine explosion, Dearborn was in a foul mood. The British and Canadians had suffered 150 killed and wounded and some 300 militiamen were prisoners, but American casualties numbered more than twice that—20 percent of his force. Dearborn reluctantly and somewhat belatedly agreed to surrender terms that returned the Canadian militiamen to their homes and promised some measure of protection for private property that fell far short of Pike's words on the subject.

And what of Chauncey's much sought-after ships? York's harbor and shipyard were not as full as reports had predicted. The unfinished *Sir Isaac Brock* was burned on its ways by the Americans, and only the rotting hulk of the six-gun *Duke of Gloucester*, laying unrigged and in the midst of repair, was able to be towed off. Some provisions and armaments were seized, but the most valuable had been blown sky-high in the blast that killed General Pike.

Then things got ugly. Despite orders to the contrary, some American troops began to loot private homes and some public buildings, including St. James Church and the town library. One American officer brashly displayed a silver service and said that it was his payback because the British had taken his property when they raided Ogdensburg, New York, the previous February. Then smoke appeared from the Parliament buildings. Constructed in 1796 when York won out over Kingston as the capital of Upper Canada, the buildings were not terribly grand. Planned as two forty-by-twenty-foot wings, they had yet to be joined by the construction of the central structure. Still, they housed the legislature, courts, and administrative offices of the province and were the only buildings in town with brick walls.

Who started the fire remains a mystery, although there are ample theories. One story is that roving American sailors discovered a scalp—long the symbol of frontier atrocities by the British and their Indian allies—hanging above the parliamentary mace in the customary place for the speaker's wig. It seems doubtful that it was put there in defiance. Ridiculously enough, it may have been put there as a joke on the speaker before the Americans even landed. No matter, it was hardly cause for putting the building to the torch. But both buildings burned.

The following day, Dearborn ordered the military buildings and Government House, General Sheaffe's residence, deliberately burned. When the American troops finally withdrew a week later and sailed for the Niagara frontier, Dearborn made off with both the parliamentary mace and General Sheaffe's ornate snuffbox. Little did he know that the British would seek revenge for his actions with a vengeance some sixteen months later.[15]

The public spin of the day was that the raid on York had been a great American victory—the first on land of the war. But Secretary of War Armstrong was privately furious with Dearborn. Armstrong wrote the general what amounted to a censure and held him accountable for Pike's death, the failure to locate the ships that Chauncey had prized, and the escape of Sheaffe's regulars. Next time, Armstrong insisted, Dearborn should conduct his campaign so that the British force was not "permitted to escape to-day that it may fight us to-morrow."[16]

And that tomorrow came more quickly than either Dearborn or Chauncey had foreseen. While Chauncey's ships were busy ferrying the remainder of Dearborn's troops to Fort Niagara and conducting a bombardment of Fort George in preparation for an attack, the British at Kingston decided to pounce on Sackets Harbor as a diversion. On May 29, 1813, some 750 British troops landed just west of town and moved to destroy Chauncey's booming shipyard and the garrison at Fort Tompkins. They did so, however, without reckoning on Jacob Brown. Commissioned as a brigadier general in the New York state militia, Brown quickly gathered four hundred regulars and five hundred militia and spread them out in the forest surrounding the British beachhead—exactly the maneuver that Sheaffe was too late in performing before Pike's troops at York. The Americans poured a steady stream of fire into the landing forces. "I do not exaggerate," wrote one British soldier afterward, "when I tell you that shot, both grape and musket, flew like hail."[17]

Even though the British quickly retreated to their ships, there was another similarity with the action at York. Expecting Brown to be defeated, a U.S. Navy lieutenant took it upon himself to blow up stores and set fire to several buildings in the shipyard. When Brown learned of the error, he was apoplectic. Brown bitterly complained to Dearborn that the error was "as infamous a transaction as ever occurred among military men."[18] Apparently Brown was not aware that "Granny" Dearborn had been involved in a few infamous transactions of his own. For his part in the stalwart defense, Brown was offered a regular army commission as a brigadier general and eventually went on to become commanding general of the U.S. Army during the presidencies of James Monroe and John Quincy Adams.

So what was "Granny" Dearborn up to now? With the cork of Kingston still completely out of the bottle of the Great Lakes, Dearborn was advancing on Fort George at the head of the lake. With him was Colonel Winfield Scott, would-be hero of Queenston Heights and of late a British prisoner of war. How was it that he was back in the field, not only now in command of the Second Artillery Regiment, but also serving as Dearborn's chief of staff?

During the War of 1812, armies could barely manage and feed their own soldiers let alone maintain prisoners, so the norm was to parole prisoners to their homes until officially exchanged. Paroled prisoners gave their word of honor—no minor event in 1812—that they would not take up arms against their captors again until officially exchanged—usually for a soldier or sailor of equal rank listed on a long roster negotiated between the respective governments. The process was lengthy, subject to political influence, and hardly very precise. Both sides understood, however, that the penalty for one who had been paroled but knowingly took up arms before he was officially exchanged was automatic death. Whether both sides could actually agree on the roster of officially exchanged and therefore free to fight again was an entirely different matter, and Colonel Scott appears to have reported to duty again under some cloud.

When Colonel George Izard, commanding officer of the Second Artillery, was selected for one of six new brigadier ranks, Scott filled his spot as colonel and the regiment's commander. Scott moved the Second Artillery from Philadelphia up the Hudson to Albany and along the Mohawk to Oswego, from where Chauncey's ships transported it westward to Fort Niagara. Scott quickly rose out of the confusion that marked Dearborn's staff, and like Pike before him, Dearborn was all too happy to turn operational matters over to Scott.[19]

In the wee hours of May 27, 1813, Scott led an advance guard of four hundred men ashore at Newark on the shores of Lake Ontario just west of Fort George and the mouth of the Niagara River. Several thousand more troops were poised to follow, while another force led by Colonel James Burn of the Second Dragoons crossed the Niagara at Queenston to block a British escape in that direction. Inside Fort George, Brigadier General John Vincent had about 1,100 British regulars and militia. Another 750 troops and 200 militia were strung along the Niagara frontier south to Fort Erie. Vincent decided to fight Scott on the beach.

Watching from the *Madison*, Dearborn saw Scott take a tumble on the slope above the beachhead and feared for a moment that he had lost another aggressive subordinate. But Scott was quickly on his feet and leading the charge inland. British regulars poured lead into the fray and stood for a time with cold steel bayonets, but gunfire from Chauncey's ships offshore eventually drove them back. Once their retreat started, there was little that Vincent could do to stem the tide, no matter how many companies he ordered into the line.

Scott hoped to sweep between Vincent's troops and Fort George. He managed to do so, but not before Vincent had evacuated the garrison and spiked its guns. From two captured prisoners, Scott learned that fuses had also been set in the fort's powder magazine. Knowing full well what had happened to General Pike, Scott rounded up two companies of the Second Artillery and galloped for the fort, arriving just as one of the magazines exploded. The concussion sent stones flying and knocked Scott from his horse, breaking his collarbone in the process, but barely slowing him down. He helped stamp out the other fuses, and Fort George was safely in American hands.

Blocked by Burn and the Second Dragoons at Queenston from reaching Fort Erie, Vincent ordered his remaining forces to retreat to the west and regroup west of Queenston. Vincent was supposed to be cut from the same cloth as Brock, but he acted more like William Hull in quickly abandoning all the posts on the Canadian side of the Niagara, including Fort Erie. At long last the Americans appeared to be getting somewhere. With the Niagara frontier taken if not yet secured, all of those cries of "On to Canada" were finally beginning to bear fruit. But what was it that Secretary of War Armstrong had told Dearborn about capturing armies and not letting them escape to fight another day?[20]

While Scott nursed his collarbone, Dearborn sent brigadier generals William H. Winder and John Chandler westward along the lakeshore in pursuit of Vincent. Winder and Chandler were two political appointees with little military experience. They proved it when Vincent counterattacked and surprised the American encampment at Stoney Creek. Winder and Chandler were both captured, and the American column retreated toward Fort George.

As it did so, the importance of naval control of the lakes once again raised its head. After Scott's landing, Chauncey's fleet had sailed east to support Brown at Sackets Harbor. The ships now firing from offshore near Fort George were flying the British ensign and aiming their bombardments at the Americans. Dearborn quickly decided to pull back all his troops to Fort George and do what he did best—fret. The Americans abandoned Fort Erie on June 9 and put a torch to it. The only good news was that in the brief American occupation of Fort Erie, the ships that had been built across the Niagara River at Black Rock sailed west and joined the fleet that Oliver Hazard Perry was building at Presque Isle.[21]

All across the Niagara peninsula on both sides of the river, an eerie and vengeful veil of guerrilla warfare descended. It was neighbor versus neighbor, friend versus friend. With the principal combatants all speaking English and indeed sometimes being related, it was difficult to know which side someone was on and which house might offer shelter or capture.

Thick in the midst of these guerrilla skirmishes was Lieutenant James Fitzgibbon and a handpicked group of fifty volunteers from Isaac Brock's old regiment, the Forty-ninth Foot. Indeed, Fitzgibbon had risen through the ranks to become an officer thanks to Brock's tutoring. Fitzgibbon established an advance post in the two-story stone house of militiaman John De Cew near the beaver dams on Twelve Mile Creek, some seventeen miles from Fort George. From there, his command—essentially a special forces unit—conducted a number of hit-and-run raids throughout the countryside.

Dearborn wearied of these attacks and dispatched Lieutenant Colonel C. G. Boerstler and a combined force of 575 cavalry and infantry along with two field guns to trounce Fitzgibbon's company and probe for a weakness in the British line that now encircled the American troops around Fort George. Boerstler moved south from Fort George to Queenston with as much secrecy as a force that size might allow. Now it was the Canadians who were about to get a hero—or in this case, a heroine.

Laura Secord was a thirty-eight-year-old Queenston housewife and mother of five. Her husband, James, was bedridden from wounds received as a militiaman with General Brock on Queenston Heights the year before. There is almost a different version of what happened next for every telling. Somehow, someway, Laura Secord learned of Boerstler's proposed attack on Fitzgibbon. On the evening of June 22, 1813, she left Queenston on foot ahead of the approaching American troops and walked almost twenty miles through black swamps and the dark of night to warn Fitzgibbon about the coming attack. Great Britain's Caughnawaga Indian allies captured her as she made her way past their camp, and they finally brought her before Fitzgibbon.

Fitzgibbon was skeptical. Who was this woman? Still, he alerted both his own troops and those of his Indian allies. When no attack came on the twenty-third, Laura Secord was questioned at length about her Loyalist upbringing and the veracity of her story. But soon Captain Dominique Ducharme, a Frenchman commanding British Indian allies, also raised the alarm that the Americans were near. Three hundred Caughnawaga soon attacked Boerstler's rear guard, while another one hundred Mohawk also joined the fray. Fitzgibbon advanced with his company, but later reported that "not a shot was fired on our side by any but the Indians. They beat the American detachment into a state of terror, and the only share I claim is taking advantage of a favorable moment to offer them [the Americans] protection from the tomahawk and the scalping knife."[22] Boerstler quickly surrendered his entire command at what came to be called the Battle of the Beaver Dams. Despite Fitzgibbon's modesty, another officer later observed that the Caughnawaga fought the battle, the Mohawk got the plunder, and Fitzgibbbon got the credit.

But Fitzgibbon's fame was minuscule compared to the legend that grew up around Laura Secord's walk. In her later years—she lived until 1868—her story was elaborated to the point that some versions include details of her driving a milk cow part of the way in an effort to conceal the real purpose behind her walk. And there are questions of how Secord learned about the pending attack. Some, including Canadian historian Pierre Berton, have suggested that given the time frame, there is no way that she could have learned of the attack solely on the basis of Boerstler's advance into Queenston. By then she was already in Fitzgibbon's camp. Did Laura Secord have a secret informant? How much did she do to protect her invalid husband and children? No matter how many details became embellished about certain parts of her walk, Laura Secord always remained tight-lipped about this one—and perhaps most important—aspect.[23]

Along the Niagara frontier, Boerstler's surrender of yet another American command confirmed Dearborn's inability to break out of the Fort George perimeter, but it also finally sealed Dearborn's fate as a theater commander. Charles J. Ingersoll, a Republican congressman from Pennsylvania, led the charge to have Dearborn replaced. "We have deposed General Dearborn," Ingersoll reported, "who is to be removed to Albany, where he may eat sturgeon and recruit."[24] Who in the world would the Madison administration find to replace him?

Don't Give Up the Ship

When Captain William Bainbridge and USS *Constitution* sailed off to fight HMS *Java*, Master Commandant James Lawrence and the eighteen-gun USS *Hornet* were left to blockade HMS *Bonne Citoyenne* in Bahia, Brazil. This Lawrence dutifully did for the better part of January 1813 until the appearance of the seventy-four-gun ship of the line HMS *Montagu* caused him to give up the effort. Instead, Lawrence sailed *Hornet* north along the Brazilian coast in search of a more equal foe. On the afternoon of February 24, he found one at anchor in the mouth of the Demerara River in Guyana. This was the eighteen-gun brig HMS *Espiegle*, but as *Hornet* maneuvered to get at her, another sail appeared. Lawrence stood back to sea to ascertain the newcomer's identity. She proved to be the eighteen-gun brig HMS *Peacock* under the command of Captain William Peake. Proud as one and the model of spit and polish, *Peacock* had sailed from the *Espiegle*'s anchorage that very morning.

Captain Peake quickly hoisted his colors and moved to engage. Lawrence remained cautious and held the *Hornet* close by the wind, seeking to get the weather gauge. Both ships ended up lumbering toward each other on the wind—*Hornet* on a starboard tack and *Peacock* on port. As the ships passed, each fired a thunderous broadside at close range. Both vessels quickly came about and continued firing. *Peacock* suffered most and surrendered just fourteen minutes after *Hornet*'s first broadside. Lawrence dispatched a prize crew, but the situation was hopeless. Proud *Peacock* filled quickly with water and then settled to the bottom in thirty-some feet of water.

Why *Espiegle* had not joined the fray was anyone's guess. She remained at anchor barely four miles away and plainly visible. Lawrence quickly cleared his decks in preparation for another fight, but despite the booming cannonades, *Espiegle*'s captain asserted that he knew nothing of the battle until the next day. By then, Lawrence had crammed the *Peacock*'s crew aboard *Hornet*— nearly doubling her complement of 150—and sailed for home. It was a cramped and crowded trip, but upon their arrival in New York as prisoners of war, the officers of the *Peacock* thanked Lawrence for his hospitality, going so far as to write: "We ceased to consider ourselves prisoners; and every thing that friendship could dictate was adopted by you and the officers of the *Hornet* to remedy the inconvenience we would otherwise have experienced from the unavoidable loss of the whole of our property and clothes owing to the sudden sinking of the *Peacock*." Sometimes, it was that kind of war.[1]

In the spring of 1813 James Lawrence was thirty-one years old. Born in Burlington on the Jersey shore of the Delaware River, he, too, had been bound for a career in law when the lure of white sails putting out to sea intervened. Lawrence served on the *Adams* and then found himself second in command to Stephen Decatur aboard the *Enterprise*. Like Decatur—or perhaps because of his service under him—Lawrence, too, was prone to bold behavior bordering on brashness. Lawrence studied Decatur's example when Decatur burned the *Philadelphia* under the guns of Tripoli, and Lawrence certainly strove to emulate it once he had his own command on the *Hornet*. Even before news of the defeat of the *Peacock* reached the United States, Lawrence's name was on the short list of promotions to full captain that President Madison sent to the Senate. In some respects, the promotion was a double-edged sword.[2]

As a full captain, Lawrence could expect command of a frigate. That much he coveted. But the frigate to which he was assigned turned out to be the *Chesapeake*, a vessel that still had not been able to shake the reputation of a hard-luck ship that stemmed from her fateful encounter with HMS *Leopard*. Then there was the matter of her crew. The two-year enlistments of the *Chesapeake*'s old crew had just expired and most were not returning because of a dispute over prize money due from earlier voyages. Thus, when Lawrence stepped aboard to assume command in Boston harbor on May 20, 1813, he was met by an assortment of officers and sailors who were largely new to one another and new to their ship. His orders were to proceed to the Gulf of St. Lawrence and disrupt convoys bearing troops and supplies to Canada.[3]

Outside Boston harbor, there was another ship with a well-known reputation. Remember Captain Philip Broke of HMS *Shannon*? Even ordering all the longboats of his squadron to tow his own ship had not enabled him to catch the *Constitution*. That escape and the *Constitution*'s subsequent victories had only hardened Broke's resolve to teach these Yankee upstarts a thing or two. Broke was thirty-seven and the epitome of the proudest and sternest traditions of the Royal Navy. If he was not the best frigate skipper in His Majesty's service, it was not because he did not endeavor to be. The same could be said of his ship. *Shannon*'s reputation was that of a first-class vessel with a highly competent crew that "became as pleasant to command as it was dangerous to meet."[4] Broke's orders were to cruise off Boston harbor and dispatch any merchantmen that might attempt to enter or exit. Should an American frigate present herself, so much the better.

At first glance, the *Shannon* and the *Chesapeake* appeared fairly evenly matched. The *Chesapeake* mounted fifty guns, twenty-eight long 18-pounders on her gun deck and eighteen 32-pound carronades plus an assortment of four other guns on her spar deck. These were capable of throwing a broadside weight of 542 pounds. The *Shannon* was armed with fifty-two guns, an identical twenty-eight long 18-pounders on her gun deck and sixteen 32-pound carronades and six other guns on her spar deck. Her broadside weight was 550 pounds.

Where the two ships differed significantly was in the makeup and training of their respective crews. *Shannon* carried a complement of 330 men, most of whom had served aboard her for some time. Indeed, Broke had been in command of *Shannon* since 1806, and part of the ship's sterling reputation was based on his model of incessant training and gunnery practice. Lawrence, on the other hand, commanded 379 men he scarcely knew. While most were experienced seamen, they had never sailed together as a crew, and indeed some late arrivals still had their duffels on deck when the Chesapeake sailed. Of her principal officers, only the first lieutenant had previous command experience aboard her. What in the world, then, was Lawrence's rush to get out of Boston harbor?[5]

Much has been made of the challenge that Philip Broke sent into Boston via an American prisoner for delivery to James Lawrence. Given its "courteousness, manliness, and candor," it was, in the words of Theodore Roosevelt, "the very model of what such an epistle should be." Broke requested Lawrence to "do me the favor to meet the *Shannon* with [*Chesapeake*], ship to ship, to try the fortunes of our respective flags." Just as had recently been done before the *Bonne Citoyenne*, Broke promised that no other ship would intervene and even told Lawrence of the particulars of the *Shannon*'s armaments and the disposition of other British ships.[6]

Doubtless the challenge was intended to get Lawrence's dander up. Doubtless it would have, had he received it. Before the challenge could be delivered, Lawrence ordered the *Chesapeake* to sea, knowing full well that a British frigate lurked nearby. Perhaps he thought that the only way to forge his assorted complement into a cohesive crew was to put to sea and test it in battle. Perhaps he sought to quit Boston before his new recruits were tempted to desert to other ships. Perhaps he simply underestimated the navy that opposed him. After all, the captains of the *Bonne Citoyenne* and *Espiegle* had refused to fight, and when the *Peacock* did, she was easily bested. Broke and the *Shannon*, of course, would prove an entirely different matter.

* * *

About noon on June 1, 1813, USS *Chesapeake* weighed anchor and made her way out of Boston harbor. As she cleared the Boston lighthouse, Lawrence ran up a banner proclaiming "Free Trade and Sailor's Rights," as if to inspire and unite her disparate crew. Broke saw *Chesapeake* coming and assumed his challenge had been delivered. He leisurely sailed *Shannon* eastward, falling off the wind now and then that the *Chesapeake* might catch up. If Lawrence thought better of the encounter, now was his chance to run. But he had made clear his intentions in a letter to Secretary of the Navy Jones just before sailing: "An English frigate is now in sight from our deck. . . . I am in hopes to give a good account of her before night."[7]

Confidently, Broke allowed the *Chesapeake* to close, even permitting Lawrence to choose the weather gauge. Only when it appeared that *Chesapeake* might pass under *Shannon*'s stern and rake her, did Broke become alarmed. But Lawrence for whatever reason chose instead to pull up within fifty yards of *Shannon*'s starboard quarter and run parallel to the British frigate. Aboard *Shannon*, Broke instructed the captain of the fourteenth gun (the eighteen-pounder closest to the stern) not to fire until he could sight into the *Chesapeake*'s forward gun ports. When *Chesapeake* drew abreast to that point, the fourteenth gun roared, and from aft forward the *Shannon*'s other guns followed suit. Lawrence answered with a broadside, and the fight was on.

Chesapeake was doing damage, but at some point Lawrence must have realized that he was up against a stouter foe than the *Peacock*. *Shannon*'s guns wreaked havoc on his deck, and one helmsman after another fell, as did many of his officers. When *Chesapeake*'s jib sail and the ties to her foretop sail were shot away, it forced her into the wind and allowed Broke to pass under her stern in the movement he had feared in reverse minutes before. This allowed *Shannon*'s broadsides to pummel *Chesapeake*'s stern and aft gun stations.

Now both ships came together, and the fluke of the *Shannon*'s anchor caught in the American frigate. Broke calmly ordered the ships lashed together and prepared to lead a boarding party himself. Lawrence figured to do the same and gave the command "Boarders away!" This action would be decided with cutlass and pistol on one deck or the other—not with long guns at one hundred yards.[8]

Within seconds of issuing the call for boarders, Lawrence fell mortally wounded. As he was carried belowdecks to his death, Captain James Lawrence uttered the words that would make him immortal and become the watchword of the American navy for centuries to come. With thoughts of Commodore James Barron and another day on the deck of the *Chesapeake* no doubt flashing through his brain, Lawrence cried out: "Don't give up the ship! Fight her till she sinks."[9]

Lawrence's order to prepare to board was not repeated by the *Chesapeake*'s bugler. He cowered, "too frightened to sound a note." Only a portion of the gun crews heard the command and fell back from their cannon to get their side arms. To the British, their haste to do so looked like flight. Broke ordered his men forward and was himself the first to step from the *Shannon*'s "gangway rail on to the muzzle of the *Chesapeake*'s aftermost carronade, and thence over the bulwark on to her quarter-deck, followed by about 20 men."[10]

Not only was there confusion among the *Chesapeake*'s disjointed crew, but there was hardly an officer left standing to lead it. Broke was met on the quarterdeck by the *Chesapeake*'s chaplain, who fired a pistol at him before Broke's broad cutlass nearly severed the man's arm. "The enemy," Broke later testified, "made a desperate but disorderly resistance."[11]

As the British moved forward from the quarterdeck, the *Chesapeake*'s second lieutenant ordered her foresail set to attempt to pull the ship away from the *Shannon* and deny Broke reinforcements. This officer, too, fell wounded, and within five minutes of Broke stepping onto her decks, all resistance aboard the *Chesapeake* came to an end. Some would later build on Lawrence's dying declaration and say that the *Chesapeake* never struck her colors. Closer to the truth is the fact that no American officer remained alive above deck to do so, and with the British in possession of the ship, it was a British officer who hauled down the Stars and Stripes and raised the Union Jack.[12]

The British were ecstatic. "Captain Broke and his crew," opined the *London Morning Chronicle*, ". . . have vindicated the character of the British Navy."[13] Philip Broke was dubbed a baronet, given the key to the city of London, and made the toast of the empire—much as had Isaac Hull after the *Constitution* dispatched the *Guerrière*. And much like that encounter, *Shannon*'s victory over the *Chesapeake* had political and emotional ramifications far beyond the defeat of one lone frigate. After suffering the humiliations of the *Guerrière*, *Macedonian*, and *Java*, the Royal Navy could savor its first defeat of an American frigate. Perhaps Britannia still ruled the waves after all.

Attesting to the fact that it was the deck action that had won the day, both ships were in relatively good shape. In fact, the *Shannon* appears to have suffered more damage to her hull than the *Chesapeake*. Neither ship lost a mast.[14] Rather than rely on tactics and gunnery—both of which Captain Broke had honed to a razor's edge—Lawrence had chosen to close with his enemy and slug it out at close range. That decision made, he had lost.

The *Chesapeake* was taken first to Halifax and then to England, where she was broken up. Lawrence received a hero's funeral. The American people chose to remember his victory over the *Peacock* rather than his rash advance on the *Shannon*. Far more important than either engagement, of course, would be the embedding in the American psyche of his dying words. "Don't give up the ship" replaced "Free trade and sailors' rights" as the battle cry of the maritime war and became the motto of the young American navy. The resolve of that statement would soon show itself on the waters of Lake Erie only a few weeks hence.

WE HAVE MET THE ENEMY

News of the *Chesapeake*'s defeat and the death of James Lawrence reached Presque Isle, Pennsylvania, on July 12, 1813. The little harbor about which Daniel Dobbins had bragged was a busy place. Finishing touches were being put on two newly constructed brigs and an assortment of smaller vessels. Their commander and the commodore of the American Lake Erie fleet was Master Commandant* Oliver Hazard Perry. Twenty-seven-year-old Perry had been a friend of Lawrence from their days commanding gunboat squadrons on the East Coast. Upon hearing the news of Lawrence's death, Perry ordered the flags on all ships to half-mast and directed his officers to wear black mourning bands.

---

* Like the terminology of the ships they commanded, the early-nineteenth-century ranks of naval officers could be a little confusing. If one commanded a ship—no matter what its size—he was referred to as its "captain." This operational title was much different from the *rank* of captain in the permanent grades of the United States Navy. The title "commodore" was also both a permanent rank—one grade above captain—and an operational title accorded the commander of two or more ships. Flying a commodore's pennant from the mast of one's flagship was a big deal that was zealously coveted—especially if you were only a lieutenant in permanent grade—and it caused chain-of-command problems more than once when two "commodores" arrived on the same scene. The rank of master commandant held by Perry on Lake Erie was akin to the army rank of lieutenant colonel and between that of lieutenant and captain in the navy.

He also sent for a sailmaker and asked him to craft a personal battle flag. When complete, the dark blue banner bore a simple exclamation in bold white letters: "Don't Give Up the Ship!"[1]

That exhortation fitted Oliver Hazard Perry to a T. Born in Wakefield, Rhode Island, in 1785, Perry came from a large family of sailors. He began his naval service as a midshipman aboard the USS *General Greene*, a small frigate commanded by his father. From his father, he learned something about both shipbuilding and the art of command. Advancing to lieutenant, the younger Perry went on to serve in the wars off Tripoli, direct the construction of gunboats in Newport, Rhode Island, and command the twelve-gun schooner USS *Revenge*. He "exuded command presence" and seems to have possessed an innate ability to "work a ship" and get the most out of his men.[2]

With such a personality and naval legacy, it was no wonder that gunboat duty in Newport paled beside the excitement of a command on the Great Lakes. When Isaac Chauncey ordered Perry to Lake Erie, Perry, like Chauncey, understood that the battle for the lake might be won as readily in the shipyards as on the waves. Determining to worry later about crossing the bar at the mouth of Presque Isle Bay, Perry immediately got its shipyard into high gear. When the two new brigs slid down their ways, however, Perry was off helping Winfield Scott capture Fort George. Once that amphibious action was successfully undertaken and the British were forced to abandon Fort Erie temporarily, Perry was able to return to Presque Isle with the ships from Black Rock. These were the brig *Caledonia* that Jesse Duncan Elliott had captured the previous fall; the converted schooners *Somers*, *Tigress*, and *Ohio*; and the sloop *Trippe*.[3]

The Americans, of course, were not alone in their awareness of the critical importance of controlling the Great Lakes. British ships had sealed Hull's fate at Detroit the year before, and British commanders knew that they represented the balance of power along the entire Northwest frontier. "The enemy," reported Isaac Brock to Governor-General Sir George Prevost, commander of all British forces in Canada, just two days before Brock's death at Queenston Heights, "is making every exertion to gain a naval Superiority on both Lakes which if they accomplish I do not see how we can retain the Country."[4]

Like the Americans, the British focused their attention and the bulk of their men and materials on Lake Ontario. Lake Erie got the dregs. Thus, when another twenty-seven-year-old lieutenant, Robert Heriot Barclay, arrived at Amherstburg, Ontario, to take command of British naval forces on Lake Erie, he did so with only three junior lieutenants, a purser, a surgeon, a master's mate, and nineteen men. But Barclay was cut from much the same cloth as Perry. Commended for his action at Trafalgar, he later lost an arm while serving aboard the frigate HMS *Diana* in the English Channel. When Barclay was posted to North America in 1812, it was with the expectation that he would soon be promoted to commander. But a year later when Royal Navy captain Sir James Yeo, overall commander of British forces on the Great Lakes, tapped him for command on Lake Erie—not, it was said, Yeo's first choice—Barclay was accorded only the temporary rank of commander.[5]

Such temporary status was perhaps understandable given the support Barclay was to receive from his superiors. "I repeat to you what I have already said to General Procter," Sir George Prevost wrote to Barclay, "that you must endeavour to obtain your Ordnance and Naval Stores from the Enemy."[6]

Actually, it wasn't quite that bad. While Perry was building his twin brigs at Presque Isle, the British were constructing at Amherstburg what was supposed to be the super-ship of the lake. Christened *Detroit* to commemorate Brock's victory and make amends for the loss of the original ship of that name, the vessel was to be 126 feet long—15 feet longer than Perry's brigs. Meanwhile, Barclay had under his command three other lake vessels that had previously slipped down the ways of the Amherstburg shipyard. These were the brig *General Hunter*, launched in 1806 and now mounted with six guns; the ship *Queen*

Charlotte, launched in 1809 and now carrying fourteen twenty-four-pound carronades and three long guns; and the schooner *Lady Prevost*, launched in 1812 and armed with ten twelve-pound carronades and two long guns. In addition, Barclay had the services of the small schooner *Chippawa* and the sloop *Little Belt*—not to be confused with the vessel of the same name that had once been confronted by USS *President*. Both of these tiny craft mounted several guns.[7]

If the British wondered what Perry was up to, they need look no further than the American newspapers. In May the *National Intelligencer* gave a complete accounting of the goings-on at Presque Isle, announcing that three gunboats were on the water and that the two brigs were about to be launched. The paper gave details of the armaments of the gunboats and reported that the canvas had not yet arrived for the brigs. "It will be in Pittsburgh by the 25th of this month," the report continued, "therefore we cannot rely on it before the middle of June." Noting that General William Henry Harrison was not expected to move north before command of the lake was secured, the paper concluded, "we are apprehensive, what with one delay after another, it will be fall before he can move against Detroit."[8] There it was—Perry's status in black and white. The British hardly needed spies when newspapers provided such complete details and flowed freely across the border.

The twin centerpieces of Perry's Presque Isle construction were indeed the sister brigs that he named after his fallen friend, James Lawrence, and his recent exploits off Niagara. *Lawrence* and *Niagara* each weighed about five hundred tons, carried two masts, and— once they arrived—square sails. They were each armed with eighteen thirty-two-pound carronades and two long twelve-pounders. Clearly, they would be most effective at close range. The remainder of Perry's fleet was a hodgepodge. There were the three schooner-rigged gunboats, *Ariel*, *Scorpion*, and *Porcupine*, that had been hastily built at Presque Isle, and the similarly reconfigured vessels that had finally been able to sail from Black Rock.

Perry had his fleet, but where he was to get the crews to man it was an entirely different matter. Isaac Chauncey was still in supreme command on the lakes, and his operations on Lake Ontario sucked dry the pipeline of men and matériel that trickled from the East Coast long before any but the castoffs and misfits reached Erie. In May Chauncey and Perry agreed that it would take 740 men to man the Erie fleet: 180 each for the new brigs *Lawrence* and *Niagara*, 60 for the *Caledonia*, and 40 each for the seven schooners. When Perry came up far short of that number and determined to sail from Erie regardless, Chauncey wrote Secretary of the Navy Jones that "I am at a loss to account for the change in Captain Perry's sentiments with respect to the number of men required . . . but if Captain Perry can beat the enemy with half that number, no one will feel more happy than myself."[9]

Aside from assembling some measure of a crew, Perry's first task was to get his fleet across the bar at the mouth of Presque Isle Bay. Barclay's job was simpler. All he had to do was to keep Perry there. This Barclay attempted to do through the month of July 1813. He had already missed a chance to intercept the Black Rock ships between Buffalo and Erie, but now they, too, were bottled up in Presque Isle Bay. Any attempt to cross the sandy bar at the bay's mouth, particularly with the heavier new brigs, would require the ships to be lightened of their armaments and be laborious at best. The ships would be sitting ducks. So Barclay cruised and waited. Perry sat and waited.

On July 31 the blockading British sails were gone. Much has been made of the reasons why. Folklore suggests that Acting Commander Barclay simply had a dinner date with a widow across the lake at Port Dover. Another account even has Barclay raising a dinner toast to boast that he expected to return to Erie "to find the Yankee brigs hard and fast aground on the bar at Erie when I return, in which predicament it would be but a small job to destroy them."[10] In the wake of his later conduct, either account seems supremely disingenuous. Most likely, given the small size of his vessels, he simply returned to Port Dover to reprovision. Why he chose to remove all five of his vessels is another matter.

At first, Perry thought that Barclay's disappearance might be a ruse, a trap to lure his prized brigs into indefensible positions. Deciding to gamble that it was not, Perry sent several of the lighter gunboats across the bar to form a protective screen and then went to work stripping *Lawrence* and *Niagara* of their guns and heavy equipment. Early in the season, the water covering the Presque Isle bar was about six feet deep. Now, in the first week of August, it had fallen to less than five feet. *Lawrence* and *Niagara* drew eight to nine feet even when stripped down. The bar itself was almost a mile wide. This meant that almost a mile of shallow water had to be somehow navigated between the bay and the deeper waters of the lake.

Presque Isle promoter Daniel Dobbins does not appear to have figured in this part of the story, but Perry was able to rely on a shipwright named Noah Brown. Using a system of "camels" and with Perry working right alongside him, Brown proposed to lift the *Lawrence* across the bar. The "camels" were two wooden boxes each fifty feet long, ten feet wide, and eight feet deep that were placed on each side of the ship. Filled with water, the boxes sank to the bottom. Stout logs were then run through the lower ports of the ship and placed atop the boxes. When the water was pumped out of the boxes, they rose toward the surface, lifting the ship so that it drew less water. That was the theory. In practice, it was a mess. The *Lawrence* made it just far enough to lodge fast in the mud. For a moment, Barclay's toast—if indeed he made it— seemed prophetic.

The camels were sunk again, the logs blocked higher, and even more equipment taken off the vessel. This time it worked, but ever so slowly. For two days, unarmed and but a skeleton of herself, *Lawrence* was inched out to the lake. When she finally reached deep water on the morning of August 3, a lusty cheer was raised. By midnight, her twenty guns were back on board. Now for her sister.

On August 4 *Niagara* began her journey with the same procedure. Sink the camels, insert and jack up the beams, pump out the air, and work like the devil. Then the air was punctuated by a cry of "Sail ho!" Barclay was back. The outlying gunboats exchanged a few rounds, and *Lawrence* prepared to give a good account of herself. *Niagara* appeared doomed. Then the British sails turned back into the lake. Some would call it part of "Perry's luck;" others would term it Barclay's mistake. Later, Barclay reported that through clouds and fog he had thought the entire American fleet to be across the bar and that he chose to retire to Amherstburg and await the completion of the *Detroit*. At 9:00 P.M. on the evening of August 4, 1813, Perry reported to Navy Secretary Jones, "I have great pleasure in informing you that I have succeeded after almost incredible labour and fatigue to the men, in getting all the vessels I have been able to man, over the bar."[11]

With Perry's fleet slipped loose from Presque Isle, both sides continued to struggle to find able men to man their ships. Any hopes of major reinforcements from Lake Ontario to either Barclay or Perry were without foundation. Having finally assembled major fleets on that lake, Isaac Chauncey and his British counterpart, Sir James Yeo, seemed determined not to risk them in combat. Barclay and Perry were playing for higher odds. Barclay's countrymen at Detroit would now be the ones to starve if he could not keep the lake open. Perry would have to face the wrath of William Henry Harrison—the one American commander he seems truly to have respected—if British supply lines were not closed.

While Chauncey and Yeo sailed about Lake Ontario, Chauncey dispatched Jesse Duncan Elliott and about a hundred men to augment Perry's force upon hearing that his fleet was across the bar. Perry was also able to recruit another hundred or so soldiers to serve aboard ships. This brought the American fleet strength to about 490 men. Perry gave Elliott command of the *Niagara*, while he took personal command of the *Lawrence*. How

Elliott, who once had been the supreme authority on Lake Erie, felt about being a subordinate would show itself when the cannonballs started to fly. Determined to keep the advantage, Perry's fleet weighed anchor on August 12 and sailed west to Sandusky Bay. His fleet brought General Harrison much-needed supplies and then dropped anchor in Put-in Bay in the Bass Islands, barely thirty miles from Barclay's base at Amherstburg. From here, Perry was able to watch Barclay's movements should he attempt to pass down the lake. Now Barclay was the one blockaded.

Meanwhile, Barclay was not faring any better in the manpower department. First of all, the *Detroit* had been delayed in construction by lack of shipwrights, and for a time those who were available posed a strike when not paid promptly. Months earlier, Barclay had bemoaned to Yeo that when *Detroit* was ready to sail "there is absolutely not a man to join her without unmanning another." When *Detroit* finally had her three masts and rigging in place and was ready for service on August 17, Barclay could not obtain proper armaments for her from Long Point, Ontario, because Perry was in control of Lake Erie. Consequently, Barclay determined to "run the *Detroit* with such guns as I can procure at Amherstburg."[12]

The result was that *Detroit* ended up with eight nine-pounders, six twelve-pounders, two eighteen-pounders, and three twenty-four-pounders—all were long guns except for one eighteen-pound and one twenty-four-pound carronade. Some of the pieces had been originally captured from the British during the Revolution and then recaptured at Detroit. In the words of naval historian Alfred Thayer Mahan, "A more curiously composite battery probably never was mounted. . . ."[13]

So armed, Barclay determined that he must get to Long Point nonetheless because Detroit's residents were dependent on flour from there as well as a host of other supplies. Noting Yeo's success in capturing two of Chauncey's schooners on Lake Ontario, Sir George Prevost wrote General Henry Procter at Detroit a pep talk intended for Barclay. "Yeo's experiences," Sir George observed, "should convince Barclay that he has only to dare and he will be successful."[14] Not that he needed one, but Oliver Hazard Perry had received his own pep talk from General Harrison.

As the sun rose on the morning of September 10, 1813, over Perry's anchorage at South Bass Island, the lookout on the *Lawrence* raised the cry of "Sails ho!" Barclay and his squadron, including the *Detroit*, were on their way down the lake and giving full indication of their willingness to fight. Perry wasted no time. Anchor chains rattled, sails were run up the masts, and anxious crewmen took up their stations. At first the wind was from the southwest—favoring the British and posing a problem for the Americans even to clear their anchorage. Inexplicably—those who believe in such things would later term it another example of "Perry's luck"—the wind shifted to the southeast. This not only allowed the American fleet a gentle breeze on which to sail from its anchorage, but also gave it the windward position in the encounter that was to come.[15]

While the details could be—and have been—debated at great length, the bottom line is that on that crisp September day, the ship-for-ship broadside weight of the Americans outnumbered that of the British. The key to the battle, therefore, would be to engage ship-to-ship and—because of the Americans' preponderance of carronades over long guns—get in close. This meant that from the American perspective, the principal matchups would be twenty-gun *Lawrence* versus nineteen-gun *Detroit*; twenty-gun *Niagara* versus seventeen-gun *Queen Charlotte*; and the two long twenty-four-pounders of *Caledonia* versus *Hunter's* ten much lighter long guns.[16]

When the British vessels were close enough to be identified, Barclay's order of battle proved to be a line led by *Chippawa* and followed by *Detroit*, *Hunter*, *Queen Charlotte*, *Lady Prevost*, and *Little Belt*. He ordered this arrangement, Barclay later explained, "so that each ship [*Detroit* and *Queen Charlotte*] might be supported against the superior force of the two brigs opposed to them." Perry had expected *Queen Charlotte* to precede the green *Detroit*, and he had accordingly assigned Jesse Elliott and the *Niagara* to take the van (the lead position) of his fleet. That sounded good to Elliott, who clearly had as many aspirations for a victory covered in personal glory as did Perry. But now the key British ships were reversed. Perry quickly made adjustments. While he moved ahead in *Lawrence* to engage *Detroit*, he ordered *Scorpion* and *Ariel* to the van of his own fleet to keep Barclay in *Detroit* from crossing *Lawrence*'s bow. This would also thwart *Chippawa*. Then Perry placed *Caledonia* behind *Lawrence* to deal at long range with *Hunter* and ordered Elliott and the *Niagara* to stay in line behind *Caledonia* and engage the *Queen Charlotte*. Jesse Elliott was not pleased with the change. The glory was in the van.[17]

The winds were light, and it took the fleets some time to close with each other. Finally, just before noon, the long guns on the *Detroit* opened up on the *Lawrence* at a distance of about one mile. The first shots fell harmlessly, but others struck home and did more damage than Perry had expected. With the wind, he moved to increase his speed and close with the *Detroit* as quickly as possible to make use of his carronades. As he did so, Perry ordered the ships behind *Lawrence* to close up the line and engage their designated opponents.

The schooners *Scorpion* and *Ariel* and the brig *Lawrence* led the American line obliquely toward the British column, aiming for the *Detroit*. When they were about three hundred yards apart, Perry turned parallel to the British ships to bring his carronade broadsides to bear. The angle at which the fleets converged placed the latter vessels in each line farther apart. When *Queen Charlotte*—her captain felled by an early shot from the long twenty-fours on the *Caledonia*—found that her carronades could not reach the *Niagara*, the acting captain, First Lieutenant Thomas Stokoe, quickly chose to pass *Hunter* and bring her carronades to bear on the *Lawrence*. *Niagara* clung stubbornly in line behind *Caledonia*, and the result was that Perry and the *Lawrence*

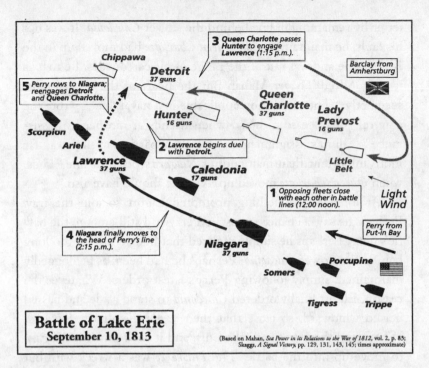

3 *Queen Charlotte passes Hunter to engage Lawrence (1:15 p.m.).*

Chippawa

Detroit
37 guns

5 *Perry rows to Niagara; reengages Detroit and Queen Charlotte.*

Barclay from Amherstburg

Queen Charlotte
37 guns

Lady Prevost
16 guns

Hunter
16 guns

Scorpion

Ariel

2 *Lawrence begins duel with Detroit.*

Lawrence
37 guns

Little Belt

Caledonia
17 guns

1 *Opposing fleets close with each other in battle lines (12:00 noon).*

Light Wind

4 *Niagara finally moves to the head of Perry's line (2:15 p.m.).*

Niagara
37 guns

Perry from Put-in Bay

Porcupine

Somers

Tigress

Trippe

Battle of Lake Erie
September 10, 1813

(Based on Mahan, *Sea Power in its Relations to the War of 1812*, vol. 2, p. 85; Skaggs, *A Signal Victory*, pp. 129, 131, 143, 145; times approximate)

were suddenly outnumbered and outgunned by the two principal British vessels.

Queen Charlotte's First Lieutenant Stokoe fell wounded, and her command devolved on an inexperienced junior officer, but the result of her maneuver to move up in line was slaughter aboard the *Lawrence*. About three hundred yards apart, the *Detroit* and the *Lawrence* blazed away at each other with broadsides, while *Queen Charlotte*'s guns had the effect of raking *Lawrence*'s stern. This lopsided duel continued for more than an hour. Barclay, though wounded in the thigh—his seventh wound during his Royal Navy career—quickly sensed that if he could disable Perry's flagship, he might win the day.

Aboard the *Lawrence*, the question that was asked with increasing frequency and increasing frustration was why didn't Jesse Elliott and the *Niagara* come to their aid? Elliott would go to his grave claiming that he had followed Perry's orders to the letter by remaining in line behind the slower *Caledonia*. It was not his fault, he maintained, that *Queen Charlotte* had moved up in the British line and was out of the range of his carronades. Be that as it may, Alfred Thayer Mahan put the heart of the matter most succinctly: "The precaution applicable in a naval duel," wrote the admiral, "may cease to be so when friends are in need of assistance." Elliott's assignment to engage a particular ship was far more important than maintaining *Niagara*'s place in line. Thus, when *Queen Charlotte* moved up, *Niagara* should have also.[18]

At long last, something prompted Elliott to join the fray. Perhaps he saw a chance to rescue Perry and still emerge the battle's hero. Perhaps he simply realized that he had tarried too long behind the slower *Caledonia*. Perhaps he had been, as he doggedly maintained, simply following Perry's latest orders. Whatever the reason, Elliott finally ordered *Caledonia* to stand aside and passed her to windward, so close that their yardarms almost touched. What wind there was had fallen off, and it took time for *Niagara* to move up into the battle. The *Lawrence* was a wreck with but one gun still operating on her starboard side. But Barclay reported that the *Detroit*, too, "was a perfect wreck," in part from the broadsides of the *Lawrence*, but also from the raking fire of *Scorpion* and *Ariel*.[19]

The situation aboard *Lawrence* was hopeless. Perry looked up into her shattered rigging and saw the blue of his battle flag. "Don't give up the ship," his friend had said while dying. "Don't give up the ship." But the ship was lost. Taking great care to leave the Stars and Stripes flying—he wanted Barclay to have no misconception that he was surrendering his squadron—Perry lowered only the banner with *Lawrence*'s dying words and determined to row to the oncoming *Niagara* to continue the fight. Draping the blue banner over his arm and leaving command of the *Lawrence* to thrice-wounded Lieutenant James Yarnall, Perry clambered down into a small cutter and with four seamen straining at its oars headed for the *Niagara*.

Aboard *Detroit*, Barclay had been wounded yet again, this

time by grapeshot that shattered the shoulder above his one remaining arm. Before he was carried belowdecks on the *Detroit*, Barclay surmised correctly what Perry was up to and directed his guns to fire at the little cutter. Cannonballs splashed around Perry's improvised flagship, and the image of him standing erect among them propelled him into legend.

When the little boat reached the windward side of the *Niagara*, Perry quickly climbed aboard and was greeted by Jesse Elliott. No accurate record remains of their conversation, but it appears to have been cordial. Elliott immediately turned command of the *Niagara* over to Perry and leaped into the cutter to row to the trailing gunboats and spur them to join the action. Seeing Perry reach the *Niagara*, Lieutenant Yarnall on the *Lawrence* struck her colors. A cheer went up aboard *Detroit* and *Queen Charlotte*, but it was short-lived as the embattled British sailors saw the fresh *Niagara* turn to engage them.

Exactly where the *Niagara* was in relation to the principal British vessels at the moment of Perry's boarding is another of the battle's controversies. Elliott's defenders maintain that he had closed with the enemy and was firing all the guns he could bring to bear when Perry boarded. Perry partisans claim the *Niagara* was out of British carronade range and still not engaged when the young commodore stepped aboard.

What happened next, however, is not in dispute. Perry ordered *Niagara* to fill her sails and turn to starboard into the British line. This maneuver brought her across the bows of the *Detroit* and *Queen Charlotte* and enabled her guns to deliver a withering raking fire to both ships. At the head of the British line, the American schooners *Scorpion* and *Ariel* kept the *Chippawa* and *Little Belt* at bay, while slow but steady *Caledonia* continued to pound away with her long guns.

Detroit turned to starboard to fight the menace on her bow and then attempted to turn completely and bring her fresh starboard battery to bear. Lieutenant Robert Irvine, the green officer who had suddenly found himself in command of *Queen Charlotte*, now came to the limits of his experience. He attempted to follow *Detroit*'s maneuver, but misjudged his turn and ran afoul of the *Detroit*'s rigging. With the two principal British ships entangled with each other, the *Niagara* poured a withering fire into both, while the long guns of the gunboats Elliott hastened to bring into action began to pound the sterns of the British vessels.[20]

The end came quickly. *Queen Charlotte* struck her colors first, followed by *Detroit*, although Barclay was below at the time. The other British vessels quickly followed suit. It was one of the very few times in the annals of the Royal Navy that an entire squadron surrendered. Of the battered *Detroit*, an observer wrote a month later that "it would be impossible to place a hand upon that broadside which had been exposed to the enemy's fire without covering some portion of a wound, either from grape, round, canister, or chain shot." The British suffered forty killed and ninety-four wounded, including Barclay, whose remaining arm would still be in a sling a year later at his court-martial. American losses were twenty-seven killed and ninety-six wounded, of which twenty-two were killed and sixty-one wounded aboard the *Lawrence*. *Niagara* had finally carried the day, but the men of the *Lawrence* had paid for it with their own blood.[21]

As the magnitude of the victory slowly sank in, Perry sat down to scribble the words that would come to be almost as famous as those on his battle flag. To General William Henry Harrison, who waited anxiously at Sandusky to learn the battle's outcome, Perry wrote: "We have met the enemy and they are ours: Two Ships, two Brigs, one Schooner, and one Sloop. Yours, with great respect and esteem, O. H. Perry." It was the epitome of understatement. To Secretary of the Navy Jones, Perry's initial report was only slightly more verbose. "It has pleased the Almighty," announced Perry, "to give to the arms of the United States a signal victory over their enemies on this Lake—The British squadron consisting of two Ships, two Brigs, one Schooner & one Sloop have this moment surrendered to the force under my command, after a Sharp conflict."[22]

• • •

Theodore Roosevelt summed up the victory succinctly and firmly. "In short, our victory was due to our heavy metal." The future president went on to write: "Captain Perry showed indomitable pluck, and readiness to adapt himself to circumstances; but his claim to fame rests much less on his actual victory than on the way in which he prepared the fleet that was to win it. Here his energy and activity deserve all praise, not only for his success in collecting sailors and vessels and in building the two brigs, but above all for the manner in which he succeeded in getting them out on the lake."[23]

But like the Battle of Tippecanoe, the Battle of Lake Erie was one of those events that would be refought over and over again in the minds of its participants and observers. On the British side, Barclay was acquitted and found to have borne himself well after inexplicably permitting Perry to cross the Presque Isle bar in the first place. On the American side, the central point of contention remained the movements of the *Niagara* prior to Perry boarding her. Why in the world had Jesse Duncan Elliott been so slow to engage the enemy? Certainly he was no coward. He had already proven his bravery long before when he snatched the original *Detroit* and the *Caledonia* from beneath the guns of Fort Erie. He proved it again after Perry's boarding *Niagara* by rowing the length of the American line to bring the rear gunboats into action to deliver the coup de grâce. Elliott steadfastly claimed that he had simply been following orders. Perry ordered him to maintain the line, and maintain the line he did, even if it meant that he could not successfully engage the *Queen Charlotte*.

It was a controversy that might have been put to rest immediately that autumn day in Put-in Bay had Perry not been so magnanimous in victory toward his subordinate. Far from criticizing Elliott, Perry praised him, going so far—according to Elliott's account—as to greet him upon his return to the *Niagara* with the words, "I owe this victory to your gallantry!" If there was any friction in that encounter, it was produced by Elliott himself, who claimed that he questioned Perry's decision to engage the enemy with the *Lawrence* before the rest of his fleet was in position. It was the officers and men of the beleaguered *Lawrence*—what was left of them—who became particularly critical of Elliott's actions. Perry chose, however, to quiet them, fearing perhaps that they might divert attention from the resounding victory. His official report noted merely that because Captain Elliott was so well known to the government, "it would almost be superfluous to speak" of him except that he evidenced his "characteristic bravery and judgment."[24]

Such damning with faint praise only fueled the controversy, even though Perry responded directly to Elliott within ten days of the battle that he found "the conduct of yourself, officers, and crew was such as to meet my warmest approbation." Perry even went so far as to convene meetings that included General Harrison to lay the matter to rest. Afterward Harrison wryly observed: "Commodore Perry has saved his [Elliott's} character for which he will never forgive him."[25]

So the Perry-Elliott controversy continued to simmer until it became ingrained in one of the most tragic rivalries within the American navy. Elliott chose to ally himself with the repudiated James Barron, another who thought that his actions had been unduly criticized or at least unappreciated. As a young officer, Perry had been one of those who joined captains Stephen Decatur and William Bainbridge in roundly criticizing Barron's actions on the *Chesapeake*.

In 1818 Elliott wrote Perry once again complaining that he had not received his fair share of glory for the victory. By now Perry was tired of being circumspect. "The reputation you lost . . . ," replied Perry caustically, "was tarnished by your own behavior on Lake Erie and has constantly been rendered more desperate by your subsequent folly. . . ." Perry went on to blame himself for screening Elliott from public censure in the first place.[26]

That was enough for Elliott. He promptly challenged Perry to a duel. Perry refused. He had just fought a duel over another matter against a marine captain, in which he held his fire while the marine missed. It took place on the same ground at Weehawken where Burr had killed Hamilton fourteen years before, and Perry's second had been Stephen Decatur. This time, in answer to Elliott's challenge, Perry chose to respond with court-martial charges. They included not only Elliott's "conduct unbecoming an officer by entering upon and pursuing a series of intrigues, designed to repair his own reputation at the expense and sacrifice of his . . . commanding officer," but also failing to "do his utmost" to come to the aid of the *Lawrence*.[27]

The Navy Department was less than pleased that two of its heroes might end up on opposite sides in a nasty court-martial. In part to delay matters, Perry was given command of a small squadron and dispatched to South America on a mission to seek restitution from the government of Simón Bolívar for the seizure of American vessels. On August 23, 1819, Perry's thirty-fourth birthday, he died of yellow fever near Port of Spain, Trinidad. Perry's court-martial charges found their way into the hands of Commodore Stephen Decatur. Before he could act on them, Decatur was killed on March 22, 1820, in a duel with James Barron that finally put to rest their long-simmering feud stemming from the *Chesapeake* incident. It was no small coincidence that Barron's second was Jesse Duncan Elliott.

Barron was finally reinstated in the navy and given command of the Philadelphia Navy Yard. Elliott lived another twenty-five years and spent much of it refighting the Battle of Lake Erie. Perry's reputation went on to surpass even that of Decatur in the annals of the United States Navy. On that autumn day off Put-in Bay, he had met the enemy and heeded the words of his friend James Lawrence not to give up the ship.[28]

OLD HICKORY HEADS SOUTH

Andrew Jackson was not one to sit things out, but for the first year of the war, that's exactly what he did—most reluctantly. At forty-five, he was the epitome of frontier Tennessee—rough, tough, impetuous, and plainspoken. Already somewhat of an elder statesman, Jackson had served the state in both the U.S. House of Representatives and the U.S. Senate and also achieved a respectable standing as a trader and businessman. Along the way, he had fought his share of battles, including a duel in which he killed his opponent over the honor of his beloved wife, Rachel. After the Battle of Tippecanoe, Jackson, as a major general of the Tennessee militia, wrote to William Henry Harrison volunteering his services. When nothing came of that, Jackson next told Tennessee governor Willie Blount that he could raise four thousand men and march on Quebec with ten days' notice. Blount passed Jackson's offer on to Secretary of War William Eustis, but received no reply.

When Congress authorized fifty thousand volunteers early in 1812, Jackson seized upon the announcement to make his own call to arms for Tennessee volunteers. Still three months before a formal declaration of war, Jackson told the young men of Tennessee that "your impatience is no longer restrained. The hour of national vengeance is now at hand." Should the government determine to invade Canada, Jackson mused, "how pleasing the prospect that would open to the young volunteer, while performing a military promenade into a distant country. . . . To view the stupendous works of nature, exemplified in the falls of Niagara . . . ," Jackson maintained, "would of themselves repay the young soldier for a march across the continent."[1]

Swept up by such grandiose rhetoric, young men flocked to Jackson's banner, and when war was finally declared, the general offered President Madison twenty-five hundred volunteers. Madison replied politely, but gave him no instructions to take the field. Jackson waited, and waited, and waited. Still no orders came. When two of Jackson's regiments were ordered north to support General Harrison, there was no assignment for Jackson. The general chafed until he realized why, and then he really chafed. Burr. It was Burr. Jackson had come too close to his intrigue, spoken too strongly in his defense, to be entrusted with a major command in a far-flung theater. Good heavens! Andy Jackson turned loose with several thousand men in the Mississippi Valley. Who knew what might happen? The Madison administration shuddered—and gave him no assignment.

Sometimes, amid the din of the war hawks' whip-poor-will cries of "Canada! Canada! Canada!" it was overlooked that many of them also had designs on Spanish Florida. Southern war hawks George M. Troup and William H. Crawford of Georgia, in particular, led the charge to acquire Florida by whatever means necessary. Some of their talk was merely the same lust for new territory that had carried Aaron Burr down the Mississippi and tens of thousands of pioneers westward across the Appalachians. But there were also definite military concerns. Disputes over the right of deposit had been settled with the purchase of Louisiana, but the American gateway of New Orleans was still precariously squeezed between Spanish interests in Florida and Texas.

Great Britain had ceded both West and East Florida (the dividing line between the two districts being the Apalachicola River) to Spain in 1783 as part of the Treaty of Paris ending the American

Revolution. Although Napoleon deposed Ferdinand VII and placed his own brother on the Spanish throne, he never managed to subdue the entire Iberian Peninsula. While Spain was nominally neutral in matters between the United States and Great Britain, the fact that the Duke of Wellington and Britannia's finest were slugging it out with Napoleon's marshals on the peninsula, in Spain's behalf, made Spain in fact a British ally. One did not have to try too hard to imagine a British army landing in Spanish territory at Pensacola or Mobile and then marching across Georgia and the Carolinas or clamping a stranglehold on the Mississippi at New Orleans. "Whoever possessed the Floridas," folks down south said, "held a pistol at the heart of the republic."[2]

While he waited impatiently for a command, Andrew Jackson fanned these sentiments by boasting of conquering the Floridas and subduing the Creek Nation in the process—exactly the sort of talk that had gotten him too closely tied to Burr in the first place. "Turn your eyes to the south," the general urged his young volunteers. "It is here [West Florida] that an employment adopted to your situation awaits your courage and your zeal" and could extend "the boundaries of the Republic to the Gulf of Mexico."[3]

Then in October 1812 Governor Blount received instructions from the War Department to dispatch fifteen hundred volunteers to New Orleans to bolster the southern frontier. The communiqué hinted strongly that Jackson's services were not among those required, but Blount was closely tied with Jackson politically and promptly inked Jackson's name on a blank commission that made him a major general of not just Tennessee volunteers, but of *United States* volunteers. Hearing that certain state militia had refused to enter Canada, Jackson wrote Secretary of War Eustis that his men had no such constitutional scruples. "If the government orders," the general declared, they would "rejoice at the opportunity of placing the American eagle on the ramparts of Mobile, Pensacola, and Fort St. Augustine, effectually banishing from the southern coasts all British influence."[4]

On January 7, 1813, Jackson moved south from Nashville.

With him were Captain William Carroll, a promising young officer from Pennsylvania, and Thomas Hart Benton, an intrepid
young lawyer who served as Jackson's aide-de-camp. Jackson led
fourteen hundred infantry in thirty-some boats down the
Cumberland, Ohio, and Mississippi rivers, while his best friend,
John Coffee, led a detachment of about six hundred cavalry overland down the Natchez Trace. The two forces rendezvoused at
Natchez, but before they could proceed on to New Orleans they
were met with a jolt. The American commander at New Orleans
requested that they proceed no farther.[5]

The American commander of the southwestern frontier—just
as he had been off and on for the better part of two decades—was
none other than Brigadier General James Wilkinson. Duplicitous
as always, Wilkinson was cordial toward Jackson, but hardly
wanted a major general of volunteers with an ostensibly independent command marching into town. Jackson's testimony at
Aaron Burr's trial had tended to paint Wilkinson the true villain,
and Wilkinson certainly had not forgotten it. Wilkinson may well
have been behind the curt War Department order that Jackson
received at Natchez on March 15, 1813, instructing him to disband his force. Certainly, Wilkinson enthusiastically enforced it.

Disband! Jackson was nearly apoplectic. If his two thousand
volunteers should suddenly be dismissed at Natchez, they would
play hell getting home to Tennessee. For his part, Wilkinson
hoped that just such a predicament might lure more than a few of
the volunteers to enlist in the regular army under his command.
Those doing so, Wilkinson assured Jackson, "would render a
most acceptable service to our Government. . . ." Wilkinson
maintained that he had no choice but to enforce the order to disband Jackson's volunteers. Wilkinson simply could not "direct the
vast expense attending the march of two thousand men in military
array across a wilderness of four hundred miles; and the more
especially as the commanding officer held himself independent of
and insubordinate to me." Having had more than his fill of
General Wilkinson and determined not to lose even one of his

young volunteers to Wilkinson or any other cause, Jackson dug into his own pocket to procure supplies for their journey back to Nashville as a unit. His Tennessee volunteers would return home together and disband together at Nashville.[6]

Scarcely had Jackson started his troops back north than General Wilkinson received authority from Congress to seize that portion of West Florida lying east of New Orleans and west of the Perdido River, the present-day boundary of Alabama and the Florida panhandle just west of Pensacola. Some claimed that this territory and its principal town of Mobile were really part and parcel of the Louisiana Purchase. "The Spanish provinces of East and West Florida," one contemporary account noted with plenty of aplomb, "having for some years past been in a revolutionary insurrectional state; and the government of Spain being unable from its embarrassments in Europe to maintain its authority over them; the American government now determined to occupy the town of Mobile, to which it had acquired a title by the purchase of Louisiana, but which still remained in the possession of the Spanish authorities."[7]

Wilkinson enthusiastically embraced such spin. Mobile was then a small town of about one hundred houses guarded by a Spanish garrison of about sixty troops at Fort Charlotte on Mobile Bay. On April 12, 1813, Wilkinson, a.k.a. Spain's "Agent 13," sent a note to the Spanish commandant, Cayetano Pérez, that Wilkinson had "come to relieve the garrison which you command from the occupancy of a post within the legitimate limits of those [United] States."[8] In other words, scram.

Pérez evacuated his post without incident and retreated with his troops to Pensacola, compliments of U.S. Navy vessels. Wilkinson promptly strengthened the fortifications at Mobile and then advanced to the Perdido River. Given the lack of resistance and his own dual role as a Spanish agent, there was not a great deal of glory in the venture, but at least he had not had to share it with Andrew Jackson. The significance of Wilkinson's actions, however, was great. No matter what trouble the British might cause with their Spanish allies at Pensacola and points in East Florida, they could no longer contemplate Mobile and its fine harbor for any offensive operations in the south, particularly against New Orleans.

Meanwhile, Jackson's Tennessee volunteers returned to Nashville, tattered and bruised but not bloodied by combat. Out of this venture, Jackson acquired the nickname "Old Hickory"—as in "tough as hickory." The exact derivation and use of the nickname varies with each telling, but what is certain is that Jackson returned to Nashville from the little foray to Natchez a revered and respected leader of men. Old Hickory was suddenly a father figure whose "benevolence, humane, and fatherly treatment to his soldiers" would, the *Nashville Whig* proclaimed, "live in the memory of his volunteers of West Tennessee."[9]

The aborted campaign set Jackson on the road to national prominence, but before he could take the first step, he was to have one last brush with the rough and tumble days of his youth. Even Jackson's principal biographer admits that the details of the affair are sketchy, but the gist of the matter is that one of Jackson's favorite young officers, Captain William Carroll, was challenged to a duel over a quarrel that had developed on the Natchez excursion. Carroll refused the challenge on the grounds that his challenger was not a gentleman. The angry challenger persisted and sent a second challenge to Carroll via Jesse Benton, the brother of Thomas Hart Benton, Jackson's aide-de-camp. When Carroll refused again on the same grounds, Jesse Benton made the challenge in his own stead, and Carroll was forced to oblige him or be branded a coward.

When young Carroll asked Jackson to be his second, Jackson wisely and politely refused, citing his older age and no doubt wishing to refrain from tarnishing his new fatherly image. But Carroll persisted, and after hearing exaggerated tales of a conspiracy to run Carroll out of the country because he held Jackson's favor, Old Hickory relented and agreed to act as Carroll's second.

The William Carroll–Jesse Benton duel was a comedy of errors. Despite his military rank, Carroll was not much of a shot. In contrast, Jesse Benton was a crack marksman who was accustomed to firing from just about any position—except, it turned out, standing erectly. The duel was held at ten paces, supposedly to compensate for the parties' difference in skills. At the command "Fire!" Jesse Benton wheeled around, dropped to a squatting position, and fired a shot that struck Carroll in the thumb. Carroll also fired and, given Benton's squatting-while-turning maneuver, managed to inflict "a long, raking wound across both cheeks of his buttocks." While perhaps stifling a laugh, Jackson was the first to say that Jesse Benton's conduct had not been "correct or honorable" because he had not stood erect after wheeling.[10]

There the matter should have ended, but it did not. Jesse's brother, Thomas, was in Washington at the time pressing Jackson's claim for reimbursement of the personal funds Jackson had expended to bring his volunteers safely home. In this Thomas Hart Benton was largely successful, only to return to Tennessee to find that Jackson's role in humiliating his brother had been blown way out of proportion. Words flew between the two men. Thomas Benton denied that he had challenged Jackson, but avowed as to how he would not shrink from a challenge from Jackson. Thomas Benton kept on talking, as did Jackson. "The trouble with Benton," Old Hickory growled, "was his inability to keep his mouth shut." Jackson promised to horsewhip Thomas the first time he saw him.[11]

All of Nashville waited for the encounter that was sure to come. Jackson's crowd hung out at the Old Nashville Inn. When the Benton brothers came to town, they registered at the City Hotel, but that was little help. One day in early September 1813, Jackson and John Coffee went strolling past the City Hotel on their way back from the post office. Thomas Benton stood in the doorway. Jackson brandished his whip, and the melee was on. Benton went for his gun. Jackson drew his and backed Benton into the hotel. From inside, Jesse Benton fired at Jackson and struck him in the arm and shoulder. As Jackson fell, he fired at Thomas, but missed. Coffee rushed to his friend's defense and fired at Thomas, but also missed. Another Jackson ally wrestled Jesse to the ground and stabbed him with a dagger. After more scuffling, all the combatants were pulled apart, and Jackson—the only one seriously hurt—was carried to the Old Nashville Inn. Amid a gush of blood he ordered doctors that he would "keep my arm." He did, but he also kept Jesse Benton's bullet in his body for twenty years. (Jackson's partisans made life miserable for the Benton boys, and they soon headed west for Missouri.) Weak from loss of blood, his shoulder shattered, and one arm in a sling, Andrew Jackson lay in a terrible condition when word reached Nashville a few days later of a horrible massacre.[12]

History's shorthand frequently suggests that the Creek Nation was goaded into attacking the United States by Tecumseh's messianic message of an Indian confederacy. There were actually many more forces at work, and in fact Tecumseh's rhetoric fell flat among a number of Creek factions as well as other southern tribes. After his August 1811 parting with William Henry Harrison at Vincennes, Tecumseh headed south certain that "the northern and southern tribes must join together in a unified front against the Long Knives."[13] But the Chickasaw were not interested in aiding tribes north of the Ohio River that had been their traditional enemies. Neither were the Choctaw, the southern tribe most closely allied with the Americans. The Choctaw certainly had no reason to carry a torch for the British, and a Choctaw chief, Pushmataha, even followed Tecumseh on his southern tour urging peace with the Americans.

Discouraged, Tecumseh headed east to his mother's people, the Creek. Among them, he met with some success. In part, this was because of a growing schism between the Lower Creek along the Chattahoochee and Flint rivers of western Georgia, who had adopted some Anglo customs and agrarian lifestyles, and the Upper Creek along the Coosa and Tallapoosa rivers of central

Alabama. The Upper Creek strongly resented the encroachments American settlers had made via a road cut through their territory to link Georgia with Mobile and New Orleans. In particular, the road had the effect of encouraging permanent settlements along the Tombigbee River north of Mobile.[14]

Add to this mix the questionable activities of the Spanish in the Floridas—both officially and unofficially—and the pot was ready to boil. Just as British influence among Indians in the northwest was a frequent point of contention, Spanish intrigue among the southern tribes—both real and imagined—led to friction among all parties. "The officers of Spain in the Floridas," one contemporary account summarized, "very amicably afforded every assistance in their power to our enemies."[15]

The warriors of the Upper Creek most closely allied with Tecumseh's dream of an Indian confederation were called Red Sticks, most probably named for the color of their Creek war clubs. A handful of Red Sticks, including Little Warrior, went north with Tecumseh and took part in the River Raisin battle. Details of how many accompanied Little Warrior and who did what upon their return vary, but the basic story is that in the course of their return south these Creek warriors murdered two white families. The American agent among the Creek demanded that the accused be turned over for trial. Instead, certain Creek leaders invoked their own justice and killed Little Warrior and three of his alleged accomplices. This action enraged the Red Sticks, who saw it as Creek subservience to the encroaching Americans, and fueled the fires of a civil war between the two Creek factions that was every bit as bitter as any animosities between the Red Sticks and the Americans.

Spanish officials in Pensacola took advantage of this turmoil by offering to arm the militant Red Sticks. Peter McQueen, a Creek leader of mixed blood, was invited to Pensacola to receive weapons and ammunition. When American settlers along the Tombigbee heard of this, they responded by raising 180 militia under Colonel James Caller and a frontiersman named Sam Dale.

On July 27, 1813, this force attacked McQueen's Creek party about eighty miles north of Pensacola as it was returning homeward. In the running skirmishes that followed, neither side could declare victory, but the Americans ended up with most of the supplies. This encounter came to be called the Battle of Burnt Corn and proved to be the opening round of the Creek War, a war that was as much Creek versus Creek as it was American versus Creek, Spanish, and British.[16]

Just as the Northwest territories became inflamed after Tippecanoe, so too did the southern frontier erupt after the Battle of Burnt Corn. American settlers quickly sought refuge in a series of hastily built stockades all along the Alabama and Tombigbee rivers. One such refuge some forty miles north of Mobile was Fort Mims. The term "fort" was stretching it a tad, but by late August 1813, the compound had seventeen buildings surrounded by a log stockade. One massive gate permitted entrance. Exactly how many people were in the structure is debatable. Estimates range from about 300 to 553, including many women and children and a force of militia under the command of Major Daniel Beasley.

Beasley was not particularly aggressive about preparing a stout defense and in fact seems to have discounted the seriousness of the Creek threat. When two slaves reported Creek painted for war in the vicinity of the fort on August 29, Beasley sent out scouts to investigate. When they returned without finding any sign, Beasley ordered the slaves whipped for spreading a false alarm. They would quickly be vindicated. Just before noon the next day, a routine drum roll summoned the community to lunch. As the men gathered inside the stockade and laid down their arms to eat, hundreds of Red Stick warriors led by Chief Red Eagle raced from the woods and stormed the gate. Belatedly, Beasley tried to close it and was among the first to fall.

The whites knew Red Eagle as William Weatherford, the son of a Scottish trader and a Creek noblewoman. Weatherford was born about 1780 and had been tutored in English. He lived in the Hickory Ground, the Creek holy ground around the confluence of the Coosa and Tallapoosa rivers. Some said that he was a reluctant warrior, but once committed he fought with a fury for the Red Stick cause. The battle at Fort Mims went on for about four hours. Creek losses appear to have been in the hundreds, but when the battle was over, almost all the fort's inhabitants—whatever their exact number—were slain. At some point William Weatherford may have tried to stop the slaughter, but his efforts were in vain. "My warriors were like famished wolves," Weatherford later explained, "and the first taste of blood made their appetites insatiable."[17]

News of the Fort Mims massacre spread quickly and galvanized the southwest. "Our settlement is overrun, and our country, I fear," warned a nearby resident, "is on the eve of being depopulated."[18] But here was ample cause, some thought as panic spread, not only to strip the Creek of their lands, but also to wrest the remainder of the Floridas from Great Britain's impotent Spanish ally. So Old Hickory—with Jesse Benton's bullet in his body, his arm in a sling, and so weak that he had to be helped onto his horse—headed south once more.

On paper, the American strategy looked bold and decisive. Four largely volunteer armies would strike to the heart of the Creek Nation and meet up at their holy ground where the Coosa and Tallapoosa rivers joined to form the Alabama. East Tennesseans under General John Cocke would march south into northern Alabama and rendezvous there with Jackson's West Tennessee boys. With Jackson in command, the combined force would continue south and converge with Georgia volunteers led by General John Floyd and the Third Regiment of U.S. Army regulars and Mississippi Territory volunteers under the command of General Ferdinand L. Claiborne, the brother of Louisiana's governor. Execution of the strategy in the field, however—just like the campaigns up north—would leave much to be desired.[19]

For starters, Andrew Jackson was never one to exhibit much patience. When General Cocke dragged his feet in moving his

East Tennessee volunteers south—he was more than a little reluctant to place them under Old Hickory's command—Jackson decided to strike independently. From a forward base at hastily built Fort Strother on the Coosa River, Jackson dispatched his trusted lieutenant, John Coffee, eastward thirteen miles to attack the Red Stick town of Tallushatchee. The action was calculated and not as precipitous as it might appear. Jackson was determined to strike a blow before the short-term enlistments of his volunteers—in some cases only sixty days—were up, and also to gain a victory with which to rally other Creek towns against William Weatherford's influences.

On November 3, 1813, Coffee approached Tallushatchee with about 900 cavalry and mounted riflemen. Encircling half of the village, Coffee lured its warriors into battle with first a feint and then a calculated retreat. The Red Sticks pursued the retreating Americans, who reached their main lines as planned and then turned to fight. Next, the ends of both of Coffee's flanks advanced and then turned to attack inward, completely surrounding the warrior force. In the end, all 186 were killed. Reported Coffee, "They made all the resistance that an overpowered soldier could do—they fought as long as one existed." The Tennesseans suffered only 5 killed and as Jackson had hoped, the action encouraged some Creek towns opposed to the Red Sticks to side with the Americans.[20]

One of those towns was Talladega, some thirty miles to the south. Scarcely had Coffee ridden back to Fort Strother than word reached Jackson that William Weatherford and a thousand Red Sticks were attacking Talladega. Its inhabitants were appealing to Jackson for aid. Where in the world were Cocke and his East Tennesseans? Jackson could surely have used them. But if Talladega fell to the Red Sticks, with it would fall the new prestige among the anti–Red Stick Creek that Coffee had won at Tallushatchee. Jackson had to respond.

Leaving only a handful of troops at Fort Strother, Jackson marched south with twelve hundred infantry and eight hundred cavalry. On November 9 Old Hickory deployed his forces around the Red Sticks surrounding Talladega and tried to emulate Coffee's tactics at Tallushatchee. Many of Weatherford's warriors rushed into the trap of the feigned retreat. The circle closed but was unable to hold, and Weatherford and seven hundred of his men escaped. That left three hundred Red Stick dead and Talladega saved. Once again, American casualties were extremely light—fifteen dead and eighty-five wounded.[21]

While Jackson returned to Fort Strother in hopes of finally joining up with Cocke, the other two prongs of the "coordinated" assault on the Creek Nation met with some success, but fell far short of the proposed grand rendezvous at the confluence of the Coosa and Tallapoosa. In late November General John Floyd and about 950 Georgia militia and 400 Cherokee allies destroyed the Creek village of Auttose on the Tallapoosa River about twenty miles east of the confluence. Two months later the Red Sticks struck back in reprisal and badly routed Floyd's command, with the result that the Georgians withdrew eastward to their home state.

The Mississippians fared only a little better. General Claiborne led his troops up the Alabama River to the confluence of its principal tributaries and on December 23 surprised William Weatherford at the village of Econochaca. Weatherford barely managed to escape, and then only by urging his horse to make a legendary and oft-recounted jump off a bluff into the river. Claiborne's troops found a note in Weatherford's house purporting to be from the Spanish governor at Pensacola congratulating the Creek chiefs on the attack on Fort Mims and assuring them of the governor's efforts to obtain additional weapons and ammunition from Havana. How much truth there was in the existence of this letter is uncertain, but reports of it had the effect of further solidifying American determination to rid the Floridas of their Spanish landlords. Before Claiborne could capitalize on his victory at Econochaca, however, lack of supplies and mass terminations of volunteer enlistments led Claiborne to march back to Mobile. That left Old Hickory pretty much alone up north at Fort Strother.[22]

In fact, Jackson was at times really alone. The short-term enlistments of many of his volunteers were also ending. Most had been ready to go home after Talledega, and many did. There is an oft-told story of Jackson himself leveling an old musket at his own volunteers and threatening to shoot the first man to desert. In the end, he relented when their enlistments were up. At one point he held Fort Strother with only 130 men. On January 14, 1814, when 800 raw recruits suddenly arrived there, Jackson wasted no time and within a week had them encamped at Emuckfaw Creek en route to an encampment the Red Sticks called Tohopeka, or Horseshoe Bend. This time the Red Sticks struck first. They attempted to attack the American line simultaneously in three places, but when one of their units failed to engage, Jackson's force held, but barely. Old Hickory retreated with a new respect for his foe and the knowledge that it would take more than 900 troops to dislodge the Creek from Horseshoe Bend.

As Jackson retreated from Emuckfaw toward Fort Strother, the Red Sticks struck again as his column was crossing Enotachopco Creek. Once again the attack was furious, but Jackson managed to rally his raw recruits and keep them from being routed. Jackson could hardly call Emuckfaw and Enotachopco strategic victories, but Red Stick losses in both engagements far outweighed his own. And the very fact that an American general had fought a campaign and retired from the field with his army intact was cause for celebration by those who remembered Hull at Detroit and the bumbling blunders along the Niagara River. Jackson's star began to shine in the eyes of those in Washington. Perhaps the Madison administration had misjudged him over the Burr brouhaha after all. A few weeks later, with new men—including the long-awaited volunteers from Eastern Tennessee and the Thirty-ninth Regiment of U.S. Infantry—arriving to take the places of those who had marched home, Jackson managed to piece together the larger force he needed to attack Horseshoe Bend.[23]

Leaving a rear guard at Fort Strother, Jackson moved south with the bulk of his army on March 14, 1814. His force numbered about four thousand, including many Lower Creek, Cherokee, and Choctaw allies. His objective was Tohopeka, the Red Stick stronghold at Horseshoe Bend, a wide, oxbow loop on the Tallapoosa River. About one thousand warriors and some three hundred women and children were fortified there on about one hundred acres of land on the bend. They were almost completely surrounded by water and had built a major breastwork of logs some five to eight feet high with many portholes across the 350-yard-wide neck of the peninsula. It was a distinctly non–Native American structure and once again some Americans saw it as a sure sign that British or Spanish advisers were coaching the Creek.

Jackson sent the stalwart John Coffee and his cavalry along with a group of Cherokee to occupy the riverbank opposite the head of the bend. This was designed both as a feint to divert attention from the main attack to come against the breastwork and as a measure to block any line of escape. At 10:30 A.M. on the morning of March 27, 1814, Jackson ordered his artillery pieces—the grand total of one six-pounder and one three-pounder—to open fire on the breastwork. It was like throwing baseballs at the side of a barn. The small cannonballs bounced off harmlessly or flew over and into the woods. The Red Sticks lifted shouts of defiance.

Then Coffee made his move. Cherokee warriors swam across the river and cut loose a number of Red Stick canoes. Returned to the opposite shore, the canoes were loaded with a small force that crossed the river at the head of the bend and set fire to some huts. This was the diversion Jackson needed, and he ordered his men to charge the breastwork. Under a hail of bullets and arrows, his soldiers welled up against the structure and poured a deadly fire through the portholes. Then Major Lemuel P. Montgomery, descendant of the fabled Richard Montgomery who had led the 1775 attack on Quebec, clambered atop the log barrier and ordered his men to follow. Montgomery fell almost instantly with a bullet in the head, but his place was taken by a young officer named Sam Houston, who repeated the command and led the way over the barricade. Soon Jackson's superiority in numbers told the story. The Red Sticks retreated under a steady hail of fire. Some tried to escape by canoe, but were cut off by Coffee's troops. Others took to the higher bluffs and dense brush. None chose to surrender.

As the fighting dragged on through the afternoon, Jackson later admitted to Rachel, "the carnage was dreadful." Old Hickory sent forward a flag of truce to end the affair, but it was met by a hail of bullets. Now Jackson leveled his two artillery pieces at the remaining Red Stick stronghold and blasted away. By the time night fell, only a few Red Sticks had escaped across the river under the cover of darkness. When Jackson counted the dead in the morning light, some 557 were found on the ground and another 300 were estimated to be dead in the river—close to 900 killed in all. There were several casualties among the women and children, but for the most part they were taken as prisoners. Jackson's own casualties amounted to 47 dead and 159 wounded, and another 23 killed and 47 wounded among his Creek and Cherokee allies.[24]

In comparison to some of the action between Americans and British on the Niagara frontier, the Red Stick losses were staggering. The power of the Red Sticks—which still might have been harnessed by the British—was smashed for good. "The fiends of the Tallapoosa," Jackson told his victorious troops, "will no longer murder our Women and Children, or disturb the quiet of our borders."[25] But their leader William Weatherford, Chief Red Eagle, had been absent from Horseshoe Bend.

Now Jackson moved south to the Creek holy ground at the junction of the Coosa and Tallapoosa rivers—just north of present-day Montgomery. Here on the ruins of an old French fort, he built Fort Jackson. The word went out to the Creek who remained in opposition: surrender and sever ties with the British and Spanish to the south or be annihilated. Many turned themselves in, but the whereabouts of William Weatherford remained a mystery. Jackson commanded the Creek to produce him to show their good faith, but before they were forced to do so, Weatherford surprised Jackson by walking boldly into his camp. It was exactly the sort of brave act guaranteed to impress Old Hickory.

"My warriors can no longer hear my voice," Red Eagle said evenly. "Their bones are at Talladega, Tallushatchee, Emuckfaw and Tohopeka. . . . I now ask for [peace] for my nation, and for myself." Weatherford set about convincing the remaining holdouts to surrender, and the Creek war came to an end.[26]

Old Hickory returned to Nashville to an outpouring of admiration that overshadowed anything that he had been accorded upon his return from Natchez the year before. Later that summer, he was back at Fort Jackson to dictate the terms of the Treaty of Fort Jackson that stripped the Creek of half their territory. "I finished the convention with the Creeks," Jackson wrote afterward to John Overton, his business partner in land speculation, and it cedes "20 million acres of the cream of the Creek Country, opening a communication from Georgia to Mobile."[27]

Such a massive land grab pleased Jackson's western supporters, but it left many aghast. By comparison, it made William Henry Harrison's acquisitions along the Wabash look like peanuts. Two decades later, as president of the United States, Jackson would preside over the further removal of the Creek as well as his Cherokee allies. Cherokee chief Junaluska, who had led five hundred Cherokee and stood with Jackson at Horseshoe Bend, said at that time: "If I had known that Jackson would drive us from our homes, I would have killed him that day at the Horseshoe."[28]

ON THE THAMES AND
ST. LAWRENCE

"**We have met** the enemy and they are ours." Waiting anxiously at Camp Seneca on the lower Sandusky River in western Ohio, General William Henry Harrison liked the sound of that. He understood what his predecessor, the disgraced General Hull, had. Lake Erie must be made secure before any successful operations could be undertaken against Detroit. But unlike Hull, Harrison had not been inclined to ignore that fact. Now, with news of Perry's victory in hand, it was Harrison's turn to advance. He was all too ready to do so after what had been a rough year.

Following the debacle at the River Raisin in January 1813, Harrison's efforts in the Northwest had been strictly defensive. Partly, this was due to the military realities of the control of Lake Erie. As long as the British dominated the lake, the lines of supply for any American thrust north from the rapids of the Maumee were, as Harrison well knew, subject to counterattacks on their flanks. But other issues were political. William Henry Harrison, for reasons not altogether clear, was not among Secretary of War John Armstrong's favorites. On March 7, 1813, the secretary's orders to Harrison decreed that the Ohio frontier would be a secondary theater of operations. Armstrong went on to restrict "Harrison's authority to call out the militia, draw supplies, or engage in offensive operations." Harrison had little choice militarily or politically but to dig in.[1]

He did so by constructing Fort Meigs at the rapids of the Maumee just upstream from the abandoned British post of Fort Miami and across the river from the site of the 1794 Battle of Fallen Timbers. Unlike many hastily built frontier posts that were fortresses in little more than name, Fort Meigs was an imposing compound. Its perimeter of twelve-foot-high pickets and dirt mounds was a staggering twenty-five hundred yards in circumference. Eight blockhouses and four artillery batteries anchored the perimeter and defended approaches to the fort from both land and water. Fort Meigs quickly became the logical forward point of deposit for men and matériel and a thorn in the side of the British should they contemplate any major advance south from Detroit.

Receiving plenty of encouragement from Tecumseh and his followers and knowing that Perry's fleet was still under construction at Presque Isle, the British commander at Detroit, General Henry Procter, determined to do just that in the spring of 1813. Mustering nine hundred regulars and militia, Procter sailed up the Maumee to the ruins of Fort Miami and with the assistance of Tecumseh and twelve hundred Indian allies proceeded to lay siege to Fort Meigs. Harrison himself was inside. Old Tippecanoe watched the British haul five artillery pieces into position for bombardment and pondered his fate. He was outnumbered almost four to one but not about to break.

"Can the citizens of a free country . . . ," Harrison exhorted his troops, "think of submitting to an army composed of mercenary soldiers, reluctant Canadians, goaded to the field by the bayonet, and of wretched, naked savages? Can the breast of an American solider, when he casts his eyes to the opposite shore [the Fallen Timbers site], the scene of his country's triumphs over the same foe, be influenced by any other feelings than hope of glory?"[2]

What Harrison's troops experienced next was not glory, but the thunder of British cannon. As it turned out, however, the projectiles they delivered landed with mere thuds. Thanks to Captain Eleazer

Wood's design of the fort and the tireless digging of Harrison's men, the British artillery rounds punched harmlessly into the dirt mounds of the fort's perimeter. Here they might have stayed, except that the Americans were decidedly short of cannonballs. Harrison promptly offered a gill (half a cup) of whiskey for each round recovered, and the digging became fast and furious. Soon the Americans were firing British cannonballs back at their attackers.

Perhaps remembering Brock's boldness before Hull at Detroit, a perplexed Procter demanded that Harrison surrender. But Old Tippecanoe was no Hull and declined emphatically. Should the post fall into Procter's hands, Harrison retorted, "it will be in a manner calculated to do him more honor . . . than any capitulation could possibly do."[3]

But Procter dallied, and soon it was his turn to have trouble with militia units. It was planting time back home, and the Canadians were eager to return to sow crops. Not used to the boredom of siege tactics, some of Tecumseh's followers drifted away as well. Then, a twelve-hundred-man relief force from Kentucky bulled its way through the British lines to reach the fort. That was enough for Procter. After skirmishing with it, he withdrew down the Maumee, and the first siege of Fort Meigs was lifted on May 9, 1813.

But Harrison was not out of the woods. Two months later, while the general still awaited the emergence of Perry's fleet from Presque Isle, Procter returned to Fort Meigs with twenty-five hundred regulars and militia and a like number of Tecumseh's warriors. This time Harrison was at Camp Seneca on the lower Sandusky, but he assured the garrison at Fort Meigs that it could withstand a second attack. Eschewing another siege, Tecumseh devised a ruse to lure the Fort Meigs garrison into the open. His warriors staged a sham battle that was supposed to be an attack on an American column arriving from Fort Stephenson on the lower Sandusky. When General Green Clay failed to take the bait and remained inside Fort Meigs, Procter and Tecumseh turned their attention to Fort Stephenson itself.

Garrisoned by only 160 men, Fort Stephenson was a fraction of the size of Fort Meigs, but would prove a tough nut to crack. Its stout stockade measured only one hundred by fifty yards but was surrounded by a moat eight feet wide and eight feet deep. A lone six-pound artillery piece could be positioned to rake the most exposed portion of the moat. But perhaps the fort's strongest feature was the fact that its twenty-one-year-old commander, Major George Croghan, was determined to hold it.

Harrison, however, had his doubts and ordered Croghan, the nephew of famed Northwest hero George Rogers Clark, to burn the post and retreat up the Sandusky to join him. Croghan responded that the general's instructions had been received too late and that "we have determined to maintain this place, and by heavens we can." Harrison took exception to the tone of the reply and initially termed it insubordination. Croghan, who had served under Harrison since Tippecanoe, was hauled before the general to account. His brash response had been partially a ruse, Croghan explained, lest the message fall into enemy hands. But if Harrison would permit him, he was certain that he could make good on it.

Begrudgingly, Old Tippecanoe agreed, and Croghan fairly flew along the muddy roads back to Fort Stephenson. He was almost too late. The next day, with gunboats in the Sandusky River and hundreds of troops and warriors surrounding the fort, General Procter demanded its surrender. Just as before the gates of Detroit the year before, the British hinted that unless the American garrison surrendered immediately, it might be impossible to restrain their Indian allies and avert a general massacre. Croghan's aide who met the proffered flag of truce responded that should the fort be taken, there would be no one left to massacre.

Scarcely had the British party returned to its lines than the gunboats and a howitzer on shore opened fire on the little fort. All through the night of August 1, 1813, the British cannon continued a steady bombardment that had little effect. Croghan replied sparingly with his lone cannon, moving it around from one position to another to give the impression that he had more than one heavy gun.

The next morning Procter launched an all-out attack. Fort Stephenson's defenders held their fire until the attackers were within a hundred yards and then unleashed a deadly volley. The advancing columns came on, clambered into the moat, and began to move along it. Suddenly Croghan's six-pounder discharged a deadly load of grapeshot the length of the moat. The British advance faltered and then fell back. Almost as quickly as they had come, the British returned to their boats and floated away down the Sandusky. Tecumseh's warriors, who Procter was quick to credit for urging the attack and just as quick to blame for its failure, melted back into the woods. "A more than adequate Sacrifice having been made to Indian Opinion," reported Procter caustically, "I drew off the brave Assailants."[4]

Tecumseh was beside himself. The warrior who had been quick to form a mutual admiration society with the bold Brock had already told Procter that he was unfit to command and should "go and put on petticoats." Now he compared the general to an animal that when frightened dropped his tail "between his legs and runs off." Such feelings did not augur well for future relations between the British commander and his principal ally when, six weeks later, General Harrison read Perry's victory report and at long last made plans to move north.[5]

Harrison had already issued a personal appeal to Kentucky governor Isaac Shelby, a hero of the Revolutionary War Battle of Kings Mountain, to take the field personally at the head of the Kentucky militia. Now Henry Clay's boast of three years before, that the militia of Kentucky alone was enough to win Canada, was about to be put to the test once and for all. Three thousand Kentuckians responded to Governor Shelby's promise that "I will lead you to the field of battle and share with you the dangers and honors of the campaign." Three times that number responded to a similar call from Ohio's governor, Return Jonathan Meigs, but Harrison politely told them that their efforts, while "truly astonishing," were unneeded. More than a little disgruntled, the Ohioans sat out what was quickly to become the recapture of Detroit.[6]

Even without the Ohioans, Harrison's army now numbered about fifty-five hundred men. Forming around Fort Meigs in late September in the wake of Perry's victory, it included twelve hundred Kentucky mounted troops under the command of Richard Mentor Johnson. Advancing to the banks of the River Raisin, Johnson's men found the mutilated bodies of their dead comrades still unburied from the slaughter of the previous winter. Among them were friends, neighbors, and relatives. The grisly scene had an effect on the Kentuckians more profound than any exhortation Johnson or Harrison himself might have delivered. Pausing only long enough to bury the remains, Johnson and his troops rode hurriedly onward toward Detroit.

Up ahead, General Procter had already seen the handwriting on the wall. "The loss of the fleet is a most calamitous circumstance," Procter confessed after learning of Barclay's defeat. "I do not see the least chance of occupying to advantage my present position, which can be so easily turned by means of the entire command of the waters here which the enemy now has, a circumstance that would render my Indian force very inefficient. It is my opinion that I should retire on the Thames without delay. . . ."[7]

Procter followed his own advice and abandoned not only Detroit, but also Fort Malden and Amherstburg on the Canadian side of the Detroit River. Any misgivings that state militia under General Hull had once had about crossing into Canada were not in evidence as the Kentuckians streamed across the river to follow. Just seventeen days after Perry's victory in the Battle of Lake Erie, Harrison occupied Amherstburg and the shipyards that had recently completed the *Detroit*.

In the face of Procter's retreat, Tecumseh remained adamant that foes were crushed by attacking, not fleeing. Procter finally convinced Tecumseh that he would stand and fight an all-or-nothing battle along the Thames River near Moraviantown some fifty miles east of Detroit. Slow and sluggish, the Thames flowed west through dense walnut forests on the peninsula between lakes Huron and Erie and was navigable for most of its length. Both armies used it as a highway for small gunboats and freight barges. In fact, Procter seems to have been more concerned with his baggage train than hastening to make a stand. It was Tecumseh who organized some rearguard action at the fords of the main tributaries of the Thames.

Finally, at dawn on the morning of October 5, 1813, Procter turned to make his stand. For some reason, he chose to do so along a wide front rather than with a concentrated force. His left flank was anchored on the Thames and ran inland 250 yards to a small swamp. Procter's own Forty-first Foot Regiment with 540 regulars and 290 men of the Royal Newfoundland Regiment were stationed there along one thin line. Tecumseh and some 500 warriors held Procter's right flank from the other edge of the small swamp to the thickets of a quite larger one several hundred yards farther inland.

Harrison arrived in the field some three hundred yards opposite the British line about 8:00 A.M. and took his time arranging his troops. With him were a handful of army regulars, Richard Mentor Johnson's mounted regiment of twelve hundred men, and another twenty-three hundred Kentucky militia. Harrison presumed that any attack directly against the British troops would open his left flank to a counterattack by Tecumseh. Accordingly, he directed Governor Shelby and the bulk of the Kentucky militia to strike Tecumseh between the two swamps, while Johnson's troops moved forward along the river to engage the British units.

But before this plan could be put into effect, scouts reported to Colonel Johnson just how thin the British line was. Johnson immediately went to Harrison with a bold plan, a frontal charge by his cavalry. It was a highly unusual frontier tactic—something decidedly more British than American—but after some thought, Harrison agreed to the daring of the plan. "The American backwoodsmen,"

Harrison later explained, "ride better in the woods than any other people. I was persuaded too that the enemy would be quite unprepared for the shock and that they could not resist it."[8]

That is exactly what happened. As the Kentucky mounted troops moved forward, someone raised the cry that soon reverberated from a thousand lips. "Remember the Raisin!" The first wave of horsemen swept through the thin British line and then dismounted and turned to surround it, pouring a steady fire on the British rear. The second wave came on to close the circle. The bulk of the two British regiments quickly surrendered, while General Procter barely managed to escape to Moraviantown. Having sent his brother, James, to lead one brigade into this part of the action, Richard Mentor Johnson led the other brigades of Kentuckians directly across the small swamp and straight into a hornet's nest of Tecumseh's warriors. Here, the fighting was much fiercer, but the American battle cry the same. "Remember the Raisin!"

In the tangled thickets at the edge of the bigger swamp, Johnson and his men were forced to dismount and fight on foot. In the thickest of the fighting, Johnson was shot through the hip and thigh. Tecumseh was also struck down in this vicinity, but his wounds proved mortal. Shelby's militia moved forward to support the mounted troops, and soon Tecumseh's remaining warriors were overwhelmed by superior numbers.

Remembering the atrocities committed on the River Raisin, the Kentuckians only made matters worse by emulating them. Tecumseh's body was mutilated and never positively identified, although there is little doubt that he died in the battle. With him died any hope for an Indian confederacy in the Northwest. Harrison himself had once said of Tecumseh that he was "one of those uncommon geniuses, which spring up occasionally to produce revolutions and overturn the established order of things."[9] One can only wonder what might have happened had it been the stalwart Brock, instead of the indecisive Procter, who stood with Tecumseh both at Fort Meigs and along the Thames.

While Harrison regrouped his forces and secured Detroit, Governor Shelby and his Kentuckians returned south, their moment of glory done. General Procter fled eastward, blaming his defeat on the inferiority of his troops—never mind that the core was his own regiment—and Tecumseh's warriors. The inevitable court-martial followed and Procter was charged with "delaying his retreat too long after the naval defeat; attempting to save useless baggage; neglecting proper preparations; failing to choose a suitable place to make a stand; and failing to exercise sufficient command during the battle." He was found guilty of failing to prepare adequately for the retreat and of making faulty tactical judgments. The court recommended suspension from duty for six months, but in the end, Procter received only a public reprimand.[10]

On its face, the Battle of the Thames was another of those little battles—total killed on both sides numbered fewer than one hundred—that had major and long-lasting ramifications. The immediate significance was that the British were chased out of Detroit and western Ontario. Harrison had driven nails into the door that Perry had slammed shut. (The British still held Mackinac Island, and an attempt to retake it that fall was foiled by the approach of winter, but with its Great Lakes artery cut, Mackinac withered on the vine and was no longer of strategic importance.) Harrison, of course, added to his Tippecanoe laurels. Perhaps more importantly, the Battle of the Thames broke the spirit of Indian resistance in the Northwest. Tecumseh was dead, and with him was gone any remaining likelihood of an Indian confederacy.

But just as with the Battle of Tippecanoe, the political ramifications of the Battle of the Thames were longer lasting. The Kentucky volunteers returned home, but in the years ahead there was no greater bond or greater compliment to be paid than to acknowledge one's presence among the walnut groves along the Thames on that crisp October morning in 1813. Many a Kentucky politician in the decades ahead owed his election to his participation, and one, Richard Mentor Johnson, rode the trail of the battle's glory all the way to the vice presidency.

Did Richard Mentor Johnson kill Tecumseh? He never claimed it directly, but neither did he do much to squelch the boasts of others that he had. Amid the smoke and confusion that fall morning, there were others who may have. But as late as 1836, when, as a congressman from Kentucky, Johnson became the vice presidential running mate of the staid Martin Van Buren of New York, his western supporters shouted, "Rumpsey dumpsey, rumpsey dumpsey, Colonel Johnson killed Tecumseh!" The Whig candidate who Van Buren beat for the presidency that year was none other than Old Tippecanoe himself, William Henry Harrison.[11]

Two far less decisive battles occurred along another river in the fall of 1813. Remember General Dearborn's belated dismissal after the summer's boondoggles at Niagara? Who did Secretary of War John Armstrong pick to replace him? Armstrong chose someone he had known since their days together as young lieutenants at the Battle of Saratoga during the Revolution, and the one man in the entire American army guaranteed to elicit almost universal feelings. Unfortunately, they were feelings of contempt. The man was finally to be made a major general after two decades as a brigadier, and his name was James Wilkinson. About the only people pleased with the appointment were folks in Louisiana— they were delighted to be rid of him on any terms.

Wilkinson for his part was not happy to leave the warmer climes. Armstrong's order to Wilkinson to report north did not reach Wilkinson in New Orleans until after his return from West Florida on May 19, 1813. Not only did Wilkinson take his time departing the Crescent City—not leaving until June 10—but he also seems to have taken his time in the journey to Washington, finally arriving there on July 31. After a round of socializing that included one of Dolley Madison's parties at the President's House, Wilkinson traveled north toward Sackets Harbor in the company of Armstrong. Not surprisingly, with the summer already more than half gone, the two spent much of the journey haggling over strategy.

Armstrong still favored an attempt on Kingston—just as he had the previous spring before Dearborn and Pike went off on their raid against York. Wilkinson was agreeable to that, provided competent forces were available to do so, but after Kingston, Wilkinson wanted to strike west into Upper Canada. Armstrong held that the greater strategic objective lay down the St. Lawrence to the east— Montreal. Such an effort, Wilkinson was quick to reply, would require the assistance of American troops around Lake Champlain commanded by Major General Wade Hampton. Would Hampton be under his command? Wilkinson asked. Armstrong assured him that he would.[12]

Wade Hampton took a different view. He was almost sixty and held no affection for James Wilkinson. Hampton had cut his teeth during the Revolution fighting alongside Thomas Sumter and Francis Marion during the guerrilla campaigns in his native South Carolina. After the Revolution he became a planter, served two terms in Congress, and then reentered the army as a colonel after the *Chesapeake* affair. In 1810 as a brigadier general, Hampton temporarily relieved Wilkinson in Louisiana when Wilkinson was summoned to Washington to account for his role in the intrigues of Aaron Burr. He would resign, Hampton told Secretary of War Armstrong, before he would take orders from Wilkinson. Only Armstrong's personal visit and his promise that all orders to Hampton would come through the War Department, and not from Wilkinson directly, placated the general to delay his resignation.[13]

Wilkinson finally reached Sackets Harbor on August 20, 1813. What he found was far from encouraging. The "competent" troops that he needed to attack Kingston were in short supply. Of the 3,483 men mustered at the garrison four days after Wilkinson's arrival, only 2,042 were reported fit for duty. Many of the others were sick with intestinal complaints that some attributed to bad flour. Truth be known, the bread dough was being mixed with water gathered from near the latrines.

Leaving a list of things to be done at Sackets Harbor, Wilkinson sailed west to Fort George to inspect the garrison there and strip it of as many troops as he dared. In the process of the six-day voyage, Wilkinson came down with some sort of ague and fever that was to plague him for the remainder of the St. Lawrence campaign. At Fort George, he found many men down with a similar malady, and more than one officer affected with the same displeasure at his arrival that had been shown by General Hampton. One was the commander of the Second Artillery Regiment, Colonel Winfield Scott, who, it will be remembered, had once been one of the general's most vocal critics.

Wilkinson certainly remembered—all too well. He instructed Colonel John Boyd and his brigade of about a thousand men to return east with him to Sackets Harbor and left Scott and his regiment at Niagara to defend what most thought would soon be a backwater. Though he had done his best to ingratiate himself to Wilkinson during the general's stay at Fort George, Scott was not pleased to be left behind. He lobbied Wilkinson incessantly and hoped to join him "in time to share in the glory of impending operations below." When news of Harrison's victory on the Thames came to Fort George, Scott used it to justify turning over command of the fort to local militia and hurrying east after Wilkinson with the regulars of his Second Artillery.[14]

By now there was more than just a little nip in the air. Winter was fast approaching. Because Commodore Isaac Chauncey had yet to engage the British Ontario fleet under Sir James Yeo in any decisive battle, Yeo was able to concentrate his ships around Kingston. This maneuver and the strengthening of the garrison there convinced Wilkinson to make only a feint at Kingston and then "slip down the St. Lawrence." His plan was to "lock up the enemy in our rear to starve or surrender, or oblige him to follow us without artillery, baggage, or provisions, or eventually to lay down arms; to sweep the St. Lawrence of armed craft, and in concert with the division under Major-General Hampton to take Montreal."[15] Duplicitous scoundrel though he might have been, no one ever termed Wilkinson a fool, and the plan on its face was a good one, but if the War of 1812 had proven one thing to date, it was that the best laid plans were rarely executed with competence.

• • •

In a roundabout way, Major General Wade Hampton was actually moving in the direction of Montreal, however begrudgingly. The obvious route was down the Richelieu River from Lake Champlain, but the British had recently taken control of the lake. The Americans had managed to hold it until June 3, 1813, when an overzealous young lieutenant named Sidney Smith sailed the eleven-gun *Eagle* and *Growler* too close to the British post on Isle-aux-Noix at the lake's northern end. Smith promptly ran his vessels aground and was soon forced to surrender. The British renamed the boats *Chub* and *Finch* and momentarily held naval superiority on the lake. Consequently, Hampton's flank would be threatened the moment he moved north. "The loss of our command on Lake Champlain at so critical a moment," lamented President Madison, "is deeply to be regretted."

So rather than descend the Richelieu, Hampton circled westward to the headwaters of the Châteauguay River and on the morning of October 21, 1813, marched into Lower Canada with four thousand infantry, two hundred dragoons, and ten field guns. In what was becoming the norm, the New York state militia refused to follow him across the border. "The perfect rawness of the troops," huffed Hampton, "has been a source of much solicitude to the best informed among us."[16]

From the border, the Châteauguay River flowed northeast about thirty-five miles before emptying into the St. Lawrence just west of Montreal. About fifteen miles above the St. Lawrence, Canadian troops under the command of Lieutenant Colonel Charles de Salaberry, a French Canadian commissioned as a regular officer in the British army, sought to make their first line of defense just upstream from a ford. On a sharp bend in the river where the Châteauguay was about forty yards wide and five or six feet deep, the defenders threw up a line of breastworks and felled a number of trees across the Americans' line of march on the west bank. De Salaberry's front line was manned by about two hundred regulars and half as many militia. In successive lines to the rear were another eleven hundred mixed troops under the command of Lieutenant Colonel George Macdonell, while three companies of militia guarded the ford on the east bank of the river.

On the afternoon of October 25, American scouts reported to Hampton on the strength of de Salaberry's front line and suggested that the position could be outflanked by moving down the east bank, crossing the ford, and surprising the Canadian rear. Colonel Robert Purdy and fifteen hundred officers and men of the First Infantry Brigade tried to do just that during the night but became hopelessly entangled in dense woods. By the time they hacked their way to the river the next morning, they were still upstream from the ford and opposite the stout defenses of de Salaberry's front line. When Purdy's advance guard finally managed to reach the ford, it came under attack from not only the militia companies stationed there, but also from the lines of Macdonell's supporting forces across the river.

Now Hampton pondered a direct frontal assault. Perhaps his days with Sumter and Marion in the swamps of South Carolina warned him about committing troops across an open field against a concealed foe. More likely, Hampton and his main force heard the racket created by Macdonell's men in the supporting ranks as they taunted Purdy's forces and yelled encouragement to their comrades on de Salaberry's front line. Drums and bugles added to the din. To Hampton, it sounded as if half the British army was amassed before him.

Without ever engaging the bulk of his command, Hampton broke off the battle and retreated south. American casualties numbered about fifty, while the Canadians lost five killed, sixteen wounded, and four missing. Grandly called the Battle of Châteauguay, it turned out to be a minor engagement hardly worthy of blunting an attack on Montreal, but that is exactly what it did. Because only Canadians were involved on the field, it also proved wishful thinking that these neighbors might not defend their borders. Of course, it further proved just how uninspired and halfhearted the entire American strategy had become. "It is the unanimous opinion of this council," wrote Hampton to Secretary of War Armstrong after meeting with his officers, "that it is necessary for the preservation of this army . . . that we immediately return by orderly marches to such a position as will secure our communications with the United States. . . ."[17] So much for Hampton supporting Wilkinson's attack down the St. Lawrence.

But Wilkinson was still headed downriver. His army of between seven thousand and eight thousand men left Sackets Harbor on October 17. Given the damp cold of the St. Lawrence Valley and relatively poor equipment and provisions, it was not in good shape. But in the worst shape of all appeared to be its general. In attempting to treat his various maladies, Wilkinson was consuming large quantities of laudanum—essentially opium. In his clearer moments, Wilkinson conceded that such treatment gave him "a giddy head." During the descent down the river, a fellow officer described him as "very merry" and reported that he sang and repeated stories.[18] Even a sympathetic biographer noted that "if he used the drug frequently, this would help explain his unstable judgment, his easy belief in enemy apparitions, and his frequent suggestions of palpably impossible schemes of campaign."[19]

It was in this frame of mind that Wilkinson welcomed Winfield Scott and his regiment to the command, apparently without pondering what consequences might befall Fort George in his absence. It may well have been the laudanum talking, but Wilkinson treated Scott cordially and placed him in the advance guard with Brigadier General Jacob Brown, the hero of the fight at Sackets Harbor the previous spring. Whether he did so out of appreciation of Scott's growing reputation as one who made things happen, or simply with a vindictive thought to put his young critic as much in harm's way as possible, is debatable. In either event, Brown and Scott made a good team as they slipped past one British force and advanced to Hoople Creek near Cornwall.[20]

But Wilkinson had left a hornet's nest in his rear. Far from his plan to "lock up the enemy in our rear to starve or surrender," some eight hundred British and Canadian troops, under the command of Lieutenant Colonel Joseph W. Morrison, were alive and well. Morrison's force had been designated a "corps of observation" to harry the American rear and report on its movements while other British units scurried to defend Montreal. Morrison performed his task so effectively that Wilkinson soon tired of this annoyance. Wilkinson turned to Brigadier General John P. Boyd and ordered him to attack Morrison with a force of about two thousand men.

On the evening of November 10, 1813, Morrison established his headquarters in the farmhouse of John Crysler on the north bank of the St. Lawrence. On his right was the St. Lawrence River. A half mile to his left was a major swamp. Between these two watery obstacles were open fields and a dirt road and fence line that offered some measure of protection. Morrison dug in and waited through a night that was filled with rain and sleet.

At dawn the next day, Boyd should have had second thoughts. Perhaps he did. Two years before at Tippecanoe, he had been the one to help Harrison establish a strong defensive position and wait to be attacked. But now on the St. Lawrence, his commanding general was no Harrison and was ordering him to attack. So under gray skies that momentarily held their rain, Boyd's troops moved cautiously back upriver to engage Morrison's force. About 2:00 P.M. on the afternoon of November 11, the American units massed in a line of trees along a field on the east edge of Crysler's Farm. Here was one last chance to postpone the encounter, but Boyd pressed onward. The center of his line, the Eleventh and Fourteenth Infantry regiments, soon stepped clear of the trees and strode forward across the field. The Americans, one British observer noted, "came on in a very gallant style."[21]

Initially there was much cheering on the American side. Reports suggested that they faced only local militia. A few rounds and determined shouts might sweep them from the field. But the opposing ranks stood silent—and still. These were not raw militia or even Canadian troops, but rather seasoned British regulars of the Eighty-ninth and Forty-ninth Foot regiments. Calm was the order of the day even in the face of erratic fire from the Americans. The British troops remained "silent and stock still firing not a shot" until the Americans had closed to between 100 and 150 yards. Then came the chilling commands from their officers: "Make ready." A short pause. "Present." Two heartbeats now. "Fire!" The result was "a withering shower of bullets" that hit the American lines with a devastating effect. One American officer reported that the British line "was the most admirable I have ever seen and its fire was in regular volleys."[22]

As the American advance faltered, Morrison's second in command suggested a charge to sweep the foe from the field. Morrison concurred, and the Eighty-ninth and Forty-ninth Foot moved forward with bayonets fixed. Boyd's forces beat a hasty retreat. "Never have so many Americans been beaten by such inferior numbers on foreign soil," bemoaned Wilkinson's biographer. But even in the midst of such firepower, casualties on both sides were relatively few. The British lost 22 killed and 148 wounded; the Americans 102 killed and 237 wounded, with another hundred or so taken prisoner.[23]

Perhaps if the eager Scott had been with Wilkinson's rear guard instead of the vanguard, things would have been different. Only Wilkinson seemed unable to admit defeat—or accept any measure of responsibility for it. Morrison's objective had been to delay Wilkinson's advance down the St. Lawrence, wrote Wilkinson to Armstrong as he scurried his bloodied rear guard downstream to follow Brown and Scott. Since Morrison had failed to do so, Wilkinson reasoned, Morrison could "lay no claim to the honours of the day." As his battered army took to boats and ran Long Sault Rapids,* Wilkinson was still insisting that since he was descending the river, "it follows incontestably that he [Morrison] had no fair ground on which to claim victory."[24]

---

* Today, like the field at Cryeler's Farm, this stretch of the St. Lawrence is inundated by the waters behind the Robert Moses Power Dam.

But Wilkinson's descent downriver toward Montreal proved short-lived. Reaching the mouth of the Salmon River just below Cornwall and now well aware that Wade Hampton's force was not moving to join him, Wilkinson turned south and went into a winter camp at French Mills on the U.S.–Canadian border. It was to prove a miserable place for any soldier unlucky enough to spend time there. Any attacks that Wilkinson launched from French Mills were purely verbal, designed to place any and all blame for the failed campaign on Hampton. On November 17 Wilkinson wrote Armstrong expressing "amazement and chagrin" at Hampton's conduct and asserting that the "game was in view, and had he [Hampton] performed the junction directed, would have been ours in eight or ten days."[25]

While Armstrong, Wilkinson, and Hampton engaged in a series of letters and intrigue designed to place the blame for another failure to capture Montreal on anyone besides themselves, the *New York Gazette* cut to the core of the matter with a little poem:

> *What fear we, the Canadians cry,*
> *What dread have we from these alarms?*
> *For sure, no danger now is nigh,*
> *'Tis only Wilkinson in arms.*[26]

Just when it looked as if things could not get any worse, the end of 1813 brought cataclysmic news from the Niagara frontier. Left by Winfield Scott to garrison Fort George, New York militia general George McClure found himself with fewer than 250 men by December. He took it upon himself to abandon the hard-won post and retreat across the river to Fort Niagara on the American side. As he did so, he also ordered the burning of the nearby Canadian village of Newark. In a blinding snowstorm and amid bitter cold, its inhabitants were put out in the streets, among them many widows and wives with small children.

Nine-year-old John Rogers watched his mother carry her cherished mantelpiece out into the street moments before their house went up in flames. Mrs. William Dickson lay ill in her bed, but was carried out, bed and all, and tossed in the snow while the thousand books of her husband's library—arguably one of the finest in Upper Canada—were turned to ashes. Mrs. Alex McKee managed to save only one item, a large tea tray that she tried to use as a sled to keep her young daughter's feet from the freezing snow—to no avail.

In all, ninety-eight houses burned that night, almost the entire town of Newark. It was hard to say who were more dazed, the four hundred refugees fleeing the town or the British troops and Canadian militia arriving on the scene. They would remember this. This was much different from the government buildings the Americans had torched at York the previous spring. This was personal. The revenge would be personal as well, and would burn far longer than the flames of Newark, not only across the river to Buffalo, but also all the way to Washington the following year.[27]

The first revenge was quick in coming. Lieutenant General Gordon Drummond had assumed command of troops in Upper Canada after Procter's defeat on the Thames, and on the night of December 18 he ordered an attack on Fort Niagara across the river from the recaptured Fort George. Despite ample notice that the burning of Newark had created a firestorm on the Canadian side, Fort Niagara was hopelessly unprepared. Some 550 British regulars and militia crossed the river, surprised the sentries at Fort Niagara's main drawbridge—which, inexcusably, was down—and captured the garrison, killing 67 and taking 350 prisoners in the process.

The British and Canadians turned next on Buffalo. General McClure fled, attributing his departure to the wrath vented by New York residents for his having started the rampage by torching Newark. Other American commanders rallied some local militia, but this force was routed at Black Rock. On December 30 the British burned the towns of Black Rock and Buffalo to the ground. In Buffalo only three buildings—the stone jail and blacksmith shop among them—were left standing. Surveying the scene a week later, Lewis Cass, veteran of the Detroit campaign, called it "a scene of distress and destruction such as I have never before witnessed."[28]

Whether General Henry "Granny" Dearborn might have done any better than James Wilkinson along the St. Lawrence and in holding Niagara that fall is doubtful, but the army had one more use for him. Dearborn served as president of the board of General Hull's long-delayed court-martial over the fall of Detroit. While Hull was found guilty as charged, Dearborn himself and most of his fellow general officers were not without indictment in the press. The *New York Gazette* took but ten lines to sum up all of the nation's failed attempts to invade Canada:

> *Pray, General Dearborn, be impartial,*
> *When President of a Court-Martial;*
> *Since Canada has not been taken,*
> *Say General Hull was much mistaken.*
> *Dearborn himself, as records say,*
> *Mistaken was, the self-same way.*
> *And Wilkinson, and Hampton, too,*
> *And Harrison, and all the crew.*
> *Strange to relate, the self-same way*
> *Have all mist-taken Canada* [sic].[29]

THE LION'S ROAR

After the defeats along the St. Lawrence and at Niagara, the winter of 1813–14 found the British-American standoff in North America little changed from the year before. True, there had been some limited American successes—the attack on York, the initial capture of Fort George, and Harrison's defeat of Procter on the Thames—but none of these engagements had been capitalized upon. Even Harrison's victory had resulted in little more than the recapture of Detroit. Any and all attempts to capture Montreal or extend the Niagara frontier had met with abysmal failure and ended—in the case of Niagara—with an all-out rout.

On the seas and waterways, laurels of victory belonged only to Perry for that September day on Lake Erie. But important though it was to retaking Detroit, it had done nothing to sever the Canadian artery of the St. Lawrence. Americans desperately looked around for some good news. Where were the *Constitution* and the *United States*, whose glorious triumphs had so thrilled the American public the year before? Bottled up. Bottled up in American harbors because of an increasingly effective British naval blockade.

The worldly superiority of the Royal Navy had, in fact, appeared to be challenged by the early victories of a few American frigates. But in reality, what the defeats of the *Guerrière*, *Macedonian*, and *Java* did most was raise the hue and cry of the British public to demand that His Majesty's government increase its naval presence in American waters and crush this upstart foe. The British Admiralty heeded the outcry, and by early 1813 there were ten ships of the line, thirty-eight frigates, and fifty-two smaller vessels operating in American waters—an advantage in capital ships of seven to one over the Americans. The noose was tightening.[1]

The Admiralty's January 1813 orders to Admiral Sir John B. Warren, commander of British ships in the North Atlantic and Caribbean, emphasized that it was "of the highest importance to the Character and interests of the Country that the Naval Force of the Enemy should be quickly and completely disposed. . . ." In support of this objective, the Admiralty had withdrawn "Ships from other important Services for the purpose of placing under your orders a force with which you cannot fail to bring the Naval War to a termination, either by the capture of the American National Vessels, or by strictly blockading them in their own Waters."[2]

With these additional ships, Warren slowly extended a blockade of the Atlantic coast north from Georgia and the Carolinas to Chesapeake and Delaware bays. By November 1813 the entire eastern coast of the United States south of New England was under a full commercial blockade. New England, while subject to military efforts such as HMS *Shannon*'s lurking off Boston harbor, was for a time immune from Warren's full commercial blockade, perhaps to reward it for its pro-British sympathies, but more realistically to allow some neutral trade between Canada and the West Indies. A commercial blockade meant that ships of any flag—even declared neutrals—attempting to cross the blockade were fair game for seizure, because, the accepted principle of international law ran, by endeavoring to defeat the efforts of one belligerent by running the blockade, they made themselves parties to the war.[3]

If American commerce had suffered during Jefferson's embargoes, the remainder of it now ground to a standstill. Such domestic trade that still occurred was slowed by the almost abysmal state of the nation's roadways. Regional shortages or surpluses fueled speculation and inflation. A hundredweight of sugar sold for $9 in New Orleans in August 1813, but commanded almost three times that amount in blockaded Baltimore.[4]

Naturally, the tightening British blockade had an equally restrictive impact on the movements of America's minuscule navy. "The British frigates," wrote Theodore Roosevelt in *The Naval War of 1812*, "hovered like hawks off every seaport that was known to harbor any fighting craft." After sinking HMS *Java*, the *Constitution* was blockaded in New England ports for most of 1813. Bold Decatur and the *United States* managed to slip from New York in May 1813, but were soon forced to seek refuge in New London, Connecticut, where they remained. The *Constellation* spent the entire war blockaded in Norfolk. *President* and *Congress* eluded the British blockade for several cruises, but *Congress* returned after eight months at sea so extensively damaged that she was stripped of her guns and laid up for the rest of the war. *Wasp* sneaked out of Portsmouth for a cruise, and then there was the *Essex*.[5]

One hundred forty feet in length and twenty-six and a half feet in the beam, the *Essex* was rated a thirty-two-gun frigate, but was crammed with forty thirty-two-pound carronades and six long twelve-pounders. Up close she was a furious fighter, but she was virtually helpless to any opponent skilled enough to hammer away at her from outside the range of her carronades. This fact caused *Essex* to be so disliked by her captain, David Porter, that he requested a transfer from the ship just prior to departing on the cruise that would make both the ship and her captain minor legends. "My insuperable dislike to Carronades and the bad sailing of the *Essex*," wrote Porter to his superiors, "render her in my opinion the worst frigate in the service."[6]

Nonetheless, on October 28, 1812, with Porter in command, *Essex* sailed from the Delaware River to rendezvous with *Constitution* and *Hornet* in the Cape Verde Islands. When the ships failed to meet, Porter took liberal interpretation of William Bainbridge's charge to employ his own discretion in such an event. Accordingly, Porter sailed *Essex* south and rounded Cape Horn, and she became the first warship to show the American flag in the Pacific Ocean. It was a bold move because much of the Pacific was still considered a Spanish lake and as such allied with Great Britain. But this also meant that it was sparsely patrolled.

Essex first called cautiously at Valparaiso, Chile, and was received cordially because that country was in revolt against Spain and considered itself a neutral. Replenished, *Essex* sailed for the Galapagos Islands and between April 17 and October 3, 1813, captured twelve British whalers. In an extraordinary show of nationalism, Porter went so far as to establish "Fort Madison" in the islands and claim several of them—despite Ecuadoran ownership—for the United States. It was a harbinger of American interest in the Pacific. Porter even put some of the islands' famous tortoises in his hold as a handy supply of fresh meat.

By the time Porter and *Essex* returned to Valparaiso early in January 1814 along with a captured vessel renamed *Essex Junior*, Porter had heard that three British warships, including the thirty-six-gun frigate HMS *Phoebe*, were searching for him. On February 8 the *Phoebe* came gliding into the neutral Chilean harbor and passed within feet of the anchored *Essex*. Both captains had their guns manned, and it has long been speculated that *Phoebe*'s captain, James Hillyar, intended to take *Essex* by surprise despite the neutrality of the port. But finding maneuvering difficult in calm winds and the *Essex* fully prepared with her short-range carronades, Hillyar quickly availed himself of such neutrality.

In the manner of the time, Porter and Hillyar exchanged pleasantries and renewed acquaintances from days in the Mediterranean. Porter proposed a single-ship duel. Hillyar refused, and several days later put to sea to blockade the port in the company of the eighteen-gun sloop HMS *Cherub*. On February 27 *Phoebe*, with *Cherub* some distance away, fired a signal gun that Porter took to mean Hillyar had had a change of heart about a ship-to-ship encounter. But when *Essex* emerged from Valparaiso harbor, *Phoebe* quickly ran downwind to join *Cherub* and refused to engage.

James Wilkinson,
patriot or scoundrel?
*National Portrait
Gallery,
Smithsonian Institution*

William Henry Harrison,
"Old Tippecanoe."
*National Portrait Gallery,
Smithsonian Institution; gift of
Mrs. Herbert Lee Pratt, Jr.*

James Madison,
fourth president
of the United States.
*National Portrait Gallery,
Smithsonian Institution*

USS *Constitution* with the wind (*right*) moves to engage HMS *Guerrière* in an oil painting by Michele F. Corne. August 19, 1812.

U. S. Naval Academy, Annapolis

USS *United States* delivers the coup de grace to HMS *Macedonian* in an oil painting by Thomas Birch. October 25, 1812.

Courtesy of the Historical Society of Pennsylvania Collection,
Atwater Kent Museum of Philadelphia

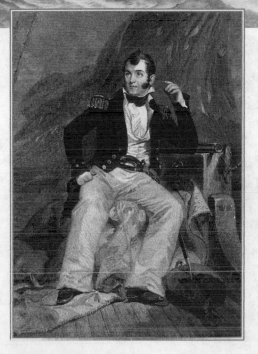

Snatching victory from defeat,
Perry is rowed from his crippled
flagship to the *Niagara*.
September 10, 1813.
Library of Congress, LC-D418-9866

Oliver Hazard Perry,
hero of Lake Erie.
Library of Congress,
LC-US762-16940

Henry Dearborn,
"Granny" to his troops.
Library of Congress,
LC-USZ62-73797

Winfield Scott,
the model hero for
a young nation.
National Portrait Gallery,
Smithsonian Institution

Thomas Macdonough,
victor of Lake Champlain.
*National Portrait Gallery,
Smithsonian Institution*

Downie sailed while Prevost dallied, and American naval forces beat the British
fleet in Plattsburgh Bay on Lake Champlain. September 11, 1814.
National Archives of Canada, C10928

Sir George Prevost,
governor-general of Canada.
National Archives of Canada,
C19123

Sir Edward Pakenham,
would-be governor
of a new province.
By courtesy of the National
Portrait Gallery, London

Sir Alexander Forester
Inglis Cochrane,
vice admiral of
the Royal Navy.
*By courtesy of the National
Portrait Gallery, London*

Sir George Cockburn,
Cochrane's torchbearer.
*By courtesy of the National
Portrait Gallery, London*

Andrew Jackson, "Old Hickory."
National Portrait Gallery, Smithsonian Institution;
gift of the A. W. Mellon Educational and Charitable Trust

Finally, on March 27, Porter determined to draw off the two British ships and thus permit *Essex Junior* to escape the harbor. Despite his earlier criticisms of his ship, Porter had seen enough of his opponents to convince him that *Essex* was fit enough to outrun both of them. For a time, it looked as if this plan would work, but just as *Essex* was rounding the outermost point of the harbor, a sudden, heavy squall struck and toppled her main topmast. Porter quickly came about to return to the shelter of the harbor, but in her crippled condition the ship was forced to anchor in a small bay three miles away. Even though *Essex* was still in the territorial waters of a neutral, *Phoebe* and *Cherub* took the occasion to pounce.

Hillyar, too, knew his opponent well, and he used his long guns to inflict significant damage on the stationary vessel while staying out of range of the *Essex's* carronades. Porter cut his anchor cable and tried to maneuver *Essex* close to his attackers, but *Essex* failed to deliver more than a couple of carronade broadsides against *Cherub* before the sloop flitted away. Next, Porter tried to run his ship aground and scuttle her. This, too, failed, and Porter was finally forced to strike his flag. Ever after he would claim that Captain Hillyar had observed the courtesies of neutrality when they served his purposes and ignored them when they did not.

From the British side, it was obvious that the Royal Navy was becoming more and more aggressive in dealing with its American antagonists. By the time news reached the United States of both *Essex's* early glories and her less than honorable demise, there was other news that promised a far more ominous turn in the war. The United States was soon to have much more to worry about than even the continued tightening of the British blockade or the reassertions of the Royal Navy's superiority.[7]

Far removed from the battles on the Thames and St. Lawrence, the face of the War of 1812 changed dramatically in mid-October 1813 in far-off Europe. Only fifteen months after launching a massive invasion of Russia at the height of his power, Napoleon met with defeat at the three-day Battle of the Nations at Leipzig. Great Britain and her continental allies breathed a collective sigh of relief. Their world order was safe. Slowly but surely over the following winter, Russian, Prussian, Austrian, and Swedish forces converged on Paris from the east. Meanwhile, Wellington's forces on the Iberian Peninsula finally expelled Napoleon's marshals from Spain and crossed the Pyrenees to invade France from the south. On March 31, 1814, these allies marched into Paris. Less than two weeks later, Napoleon abdicated all of his pretenses at empire and was exiled to Elba.*

With Europe at peace for the first time in more than a decade, the British lion was able at long last to focus its energies on North America. Now, what had those upstart Americans been doing? Back in the summer of 1812, both the *Caledonian Mercury* of Edinburgh and the *London Times* had agreed that impressment alone was too "paltry an affair for two great nations to go to war about."[8] But such sentiments had changed. British ships arriving off New England that same summer with the olive branch of Parliament's repeal of the hated Orders in Council had been seized as the first prizes of war. Next came the humiliating defeats of the *Guerrière*, *Macedonian*, and *Java*. And even if the Americans had been largely inept in execution, they had certainly tried their best to pluck Canada from the British Empire. The lion was getting annoyed. By the spring of 1814, the *London Times* sang a different tune. "Chastise the savages [Americans]," the newspaper proclaimed, "for such they are, in a much truer sense, than the followers of Tecumseh or the Prophet."[9]

At first, many Federalists celebrated Napoleon's defeat and assumed it would mean a quick reconciliation with Great Britain. Even Republican Thomas Jefferson heralded Napoleon's downfall, but other Republicans were less than optimistic about an early end to hostilities. The lion's tail had been twisted too

---

* Napoleon, of course, would escape from Elba and have one last fling at empire, but by then the British-American war in North America would be over.

sharply. "I have it much at heart," confessed the new Royal Navy commander in North America, Vice Admiral Sir Alexander Cochrane, "to give them a complete drubbing before peace is made, when I trust their northern limits will be circumscribed and the command of the Mississippi wrested from them."[10]

Having failed to conquer Canada or procure any major maritime concessions from the British in two years of war, the United States was about to spend the final year of the conflict on the defensive. The British lion was now fully engaged and greatly annoyed. The lion's roar would soon be heard in all corners of North America. "We should have to fight hereafter," wrote Albert Gallatin's brother-in-law, Joseph H. Nicholson, to Secretary of the Navy William Jones, "not for 'free Trade and sailors rights,' not for the Conquest of the Canadas, but for our national Existence."[11]

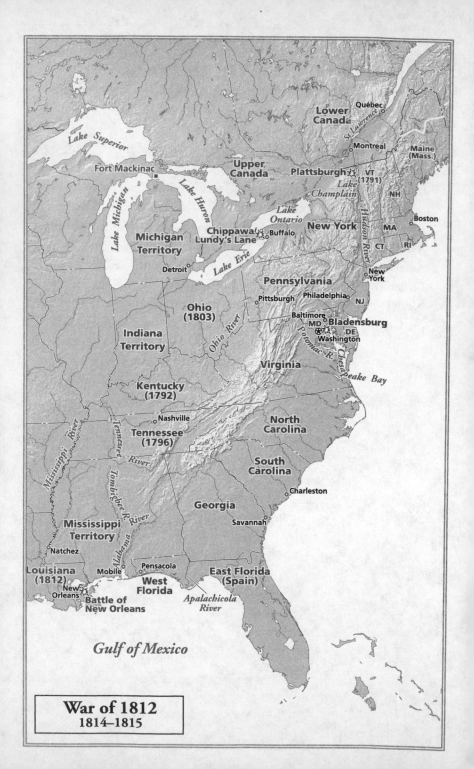

Lake Superior

Lower
Canada

Québec

Montreal

St. Lawrence

Maine
(Mass.)

Fort Mackinac

Lake Huron

Upper
Canada

Plattsburgh

VT
(1791)

Lake
Champlain

NH

Lake Michigan

Michigan
Territory

Chippawa
Lundy's Lane

Lake
Ontario

Buffalo

New York

Hudson River

MA

Boston

Lake Erie

CT

RI

Detroit

Pennsylvania

New
York

Ohio
(1803)

Pittsburgh

Philadelphia

NJ

Ohio River

Baltimore

Bladensburg

Indiana
Territory

MD

Washington

DE

Potomac R.

Virginia

Chesapeake Bay

Kentucky
(1792)

Mississippi River

Tennessee River

Nashville

Tennessee
(1796)

North
Carolina

South
Carolina

Charleston

Tombigbee R.

Georgia

Savannah

Mississippi
Territory

Alabama River

Natchez

Mobile

Pensacola

East Florida
(Spain)

Louisiana
(1812)

New
Orleans

West
Florida

Apalachicola
River

Battle of
New Orleans

Gulf of Mexico

War of 1812
1814–1815

BOOK THREE
Finale
(1814–1815)

We should have to fight hereafter not for
"free Trade and sailors rights," not for the Conquest
of the Canadas, but for our national Existence.

—*Joseph H. Nicholson to Secretary of the Navy William Jones,*
May 20, 1814

NIAGARA'S THUNDER

Despite the dismal results of the Canadian invasions of 1812 and 1813, the spring of 1814 found the American whip-poor-will still monotonously intoning, "Canada, Canada, Canada." Lake Erie remained in American hands after Perry's victory, and from its waters the Americans still hoped to wrestle at least Upper Canada from the British crown. On May 15, 1814, Lieutenant Colonel John B. Campbell led seven hundred American troops north across the lake to attack Port Dover. This was the little harbor where Robert Barclay had dallied while Perry escaped the bar at Presque Isle the summer before. It was also reported to be the haven of many of those who had put Buffalo to the torch the previous winter.

Landing at Port Dover, Campbell ordered a similar reprisal, and the result was "a scene of destruction and plunder" that one Pennsylvania soldier asserted "beggars all description." Campbell's superiors were equally aghast. Campbell was reprimanded for such excesses by a court of inquiry, but this did little to appease the British, who now had one more reckless and wanton act to avenge once their troops marched on American soil.

To the east of Lake Erie that same spring, Lake Ontario continued to be a conundrum. The British launched a raid against Fort Oswego in the lake's southeast corner in an attempt to disrupt supply lines to Sackets Harbor. Lieutenant General Gordon

Drummond's force succeeded in capturing the American post, but quickly destroyed it and withdrew without further offensive operations. Meanwhile, the opposing naval commanders on Lake Ontario, Isaac Chauncey and Sir James Yeo, remained content to parry and repose each other's halfhearted thrusts, all the while still hoping to win the naval war on the lake in the shipyards by building bigger and bigger vessels. Thus, when another American force led by the hero of Fort Stephenson, Major George Croghan, struck north from Lake Erie and failed to recapture Mackinac Island, all attention along the Great Lakes frontier centered near the falls of the Niagara.[1]

Once again the Americans prepared to cross the Niagara River. Their new commander, Major General Jacob Brown, was a decided improvement over the likes of "Granny" Dearborn and Alexander Smyth. Born in Bucks County, Pennsylvania, thirty-nine-year-old Brown represented the next generation of general officers that Secretary of War John Armstrong was rapidly pushing to the forefront. Brown had studied law in New York City, served as military secretary to Alexander Hamilton, and been made a brigadier general of the New York militia. His zealous defense of Sackets Harbor in May 1813 prompted Armstrong to offer him a commission in the regular army.

Brown emerged from Wilkinson's ill-fated march down the St. Lawrence in the fall of 1813 with his reputation intact, and Armstrong now turned to him to command the Great Lakes front of the American army. In addition to Brown, Armstrong's list of new general officers included George Izard, Winfield Scott, and Henry Clay's war hawk crony, Peter B. Porter. (The only name surprisingly absent from Armstrong's list was that of William Henry Harrison. The victor of the Thames had resigned his commission in disgust that spring after ongoing turf battles with Armstrong.) Perhaps most important, whatever else his talents, Jacob Brown was content to place Winfield Scott in a position of confidence and entrust to him much of the drilling and discipline of his army.

After the hapless engagements along the St. Lawrence in the fall of 1813, Scott had been spared the disease and drudgery of General Wilkinson's winter encampment at French Mills and been summoned to Washington. Recognizing him as one of the few rising stars in the army's still muddled officer corps, President Madison appointed Scott a brigadier general a few months shy of his twenty-eighth birthday and sent him to Buffalo as one of Brown's brigade commanders. Scott was finally in his element, and he went to extraordinary lengths to drill the four regiments of regulars in his brigade. The only thing irregular about his troops was the fact their uniforms were undyed gray and not the blue of most regular units.

The number of Scott's troops varied daily from sickness and desertion, but was generally about two thousand. "The men are healthy, sober, cheerful, and docile," wrote the youthful general. "If, of such materials, I do not make the best army now in service, by the 1st of June, I will agree to be dismissed from the service."[2] General Brown decided that the opening salvo of this new Niagara campaign would be to capture Fort Erie—something that the Americans had done once before in 1813.

Early on the morning of July 3, 1814, under the cover of a heavy rain, Winfield Scott led his brigade across the Niagara River in small boats. Nearing the shore in the lead boat, Scott probed the water with his sword and reported it only knee-deep. Always one to be in the lead, Scott promptly leaped into the water and just as promptly disappeared from sight. Somehow, he had managed to jump into a deep hole. No one laughed as the serious, young general, who always valued his military appearance, was hauled back into his boat dripping wet. Trying again, Scott found firmer ground and led his men ashore. Brigades led by Eleazar W. Ripley and Peter B. Porter soon joined them. By noon Fort Erie was surrounded, and by early evening its meager garrison of 170 men surrendered.

Downstream at Fort George, British Major General Phineas Riall heard of Fort Erie's envelopment and galloped south to try

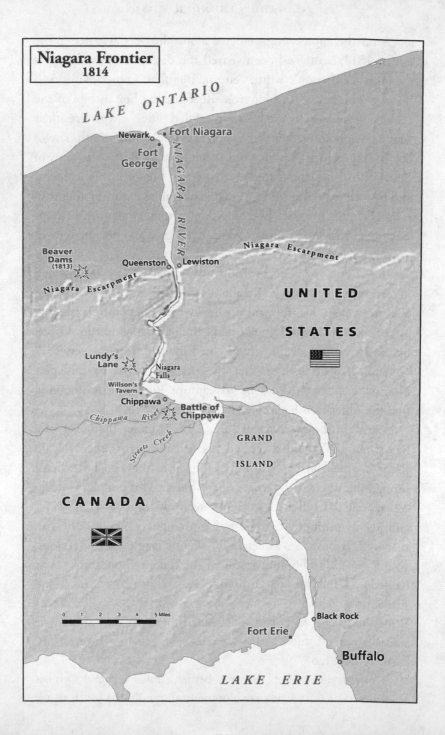

Niagara Frontier
1814

LAKE ONTARIO

Newark
Fort Niagara

Fort George

NIAGARA RIVER

Beaver Dams (1813)

Niagara Escarpment

Queenston
Lewiston

Niagara Escarpment

UNITED

STATES

Lundy's Lane

Niagara Falls

Willson's Tavern

Chippawa

Battle of Chippawa

Chippawa River

Streets Creek

GRAND

ISLAND

CANADA

0 1 2 3 4 5 Miles

Black Rock

Fort Erie

Buffalo

LAKE ERIE

to stem the American advance down the Niagara. Doubtless he carried with him thoughts of Sir Isaac Brock's similar ride from Fort George to Queenston Heights almost two years before. How would his ride end?

General Riall managed to rally British and Canadian troops and form a defensive line along the Chippawa River where it emptied into the Niagara about two miles above the falls. The little town of Chippawa at its mouth was important to the British as the southern (upstream) terminus of the Portage Road that enabled men and matériel to move around Niagara Falls. Riall ordered two companies of the 100th Foot to march south along the Niagara's western bank and form an advance guard. By coincidence, they were under the command of Lieutenant Colonel Thomas Pearson, who had been Winfield Scott's captor after the Battle of Queenston Heights.

On the morning of July 4, with exhortations to make this an Independence Day to remember, Brown ordered Scott to move downstream from Fort Erie toward Fort George. Scott did so, but quickly encountered Pearson's advance guard. Pearson's troops fought so effective a delaying action as they retreated toward Riall's main position on the Chippawa that it took the better part of the day for Scott to bring his brigade up to the river's southern bank. By now, Riall had about two thousand troops in position, and Scott wisely chose to wait for Ripley and Porter's brigades before attempting to cross the stream and engage the superior numbers. Scott went into camp on tiny Street Creek about a mile and a half south of Riall's position on the Chippawa and assumed that nothing of significance would occur in the no-man's-land in between the two armies. The boy general who had cut his teeth at Queenston Heights and come of age on the St. Lawrence still had a lesson or two to learn.

Early the next morning General Scott—never one to decline a sumptuous meal to fill his six-foot-four frame—accepted the breakfast invitation of Mrs. Samuel Street, whose property sat on Street Creek just north of the American encampment. Obviously charmed by his hostess, Scott appears to have forgotten which side of the Niagara River he was on. As he and his staff sat down to breakfast, a company of Riall's Indian allies swept out of the woods in an attempt to capture them. Mrs. Street coyly feigned surprise, while Scott and his aides scurried back to their lines, unfed but alive.

By that afternoon General Brown was making plans to flank Riall's position on the Chippawa with Ripley and Porter's brigades, and Scott's brigade moved forward across Street Creek into the supposed no-man's land. But they were not alone. Riall had boldly abandoned his defensive position on the north bank of the Chippawa and moved south to engage Brown's entire army. Scott's brigade, however, was suddenly sticking out in front. If retreat occurred to the man who had rushed up Queenston Heights, it did not show. Scott hastily hurried his remaining regiments across the narrow bridge over Street Creek and formed a battle line anchored by three artillery pieces. The British, too, had brought up artillery, and they began to pound the American position. Noting the gray of the American uniforms, Riall assumed Scott's men to be only local militia and expected the Americans to disperse quickly as his troops advanced. When that didn't happen, Riall took a closer look and realized his mistake. "Those are regulars, by God!" legend has him uttering.[3]

Both lines came on, alternately stopping to fire, reload, and move forward again. When Riall's right flank began to detach from the woods on the western edge of the field, Scott sent Major Thomas S. Jesup's Twenty-fifth Infantry regiment to turn it. As the British right wavered, Scott ordered the remainder of his troops to divide in the center and pivot inward. Riall took the bait and thought that Scott's center was folding. He ordered the Royal Scots and the 100th Foot to charge the American center. Suddenly caught in a crossfire delivered from both sides and pummeled by Scott's three artillery pieces, the British advance faltered and their surviving troops quickly retreated from the field.

This battle of a thousand or so troops on either side was hardly comparable to the gargantuan epics the British were accustomed to fighting against Napoleon. But for the Americans, after two years of stumbles and bumbles, they could at long last claim—however fleetingly—that they had bested British regulars. Scott was probably right when he wrote years later in his memoirs that "history has recorded many victories on a much larger scale than that of Chippawa; but only a few have wrought a greater change in the feelings of a nation."[4] Finally the Americans had something to celebrate. Perhaps even more important, the American public, so craving a hero on land to rank with the likes of Isaac Hull, Decatur, and Perry on water, would find one in young Winfield Scott.

Buoyed by Scott's success at the Chippawa, Brown soon moved his entire army north across the river. His plan was to link up with Isaac Chauncey's naval units on Lake Ontario before marching westward around the lake toward York. But if the American army was finally working more effectively, interservice cooperation had not improved. General Brown pleaded with Commodore Chauncey to support his advance. "I do not doubt my ability to meet the enemy in the field and to march in any direction over his country, your fleet carrying for me the necessary supplies," wrote Brown. "We can threaten Forts George and Niagara, and carry Burlington Heights and York, and proceed direct to Kingston and carry that place. For God's sake let me see you: Sir James [Yeo] will not fight."[5]

Chauncey, of course, had already spent more than a year proving that he would not fight, either. Nonetheless, the commodore caustically replied to Brown that his fleet had been created to "fight the enemy's fleet, and I shall not be diverted in my efforts to effectuate it by any sinister attempt to render us subordinate to, or an appendage of, the army."[6] Enough said. It quickly dawned on General Brown that whatever happened next, he and his army were on their own.

Brown advanced as far as the old battleground at Queenston Heights, but he was uneasy. All along his front General Riall and the British were receiving reinforcements daily, including Colonel Joseph Morrison's Eighty-ninth Foot, the regiment that had carried the day at the Battle of Crysler's Farm. Most significant was the arrival of Lieutenant General Gordon Drummond, now the supreme British commander in Upper Canada.

A Scotsman, Drummond was born in Quebec in 1772 while his father was serving there as deputy paymaster general. Entering the British army at seventeen, Drummond used the purchase system to rise rapidly in rank, commanding a regiment in Flanders alongside the future Duke of Wellington and serving in various posts throughout North America. While he lacked the personal charisma of Isaac Brock, Drummond may have been the most competent British general to follow him. After leading the May 1814 attack on Fort Oswego, Drummond waged an ongoing war of words with Canadian Governor-General Sir George Prevost about the necessity of more troops for Upper Canada before hurrying to Riall's aid after Chippawa. Drummond arrived in York on July 22, and two days later disembarked across the lake at Fort George. Whatever else Commodore Chauncey was doing, he wasn't disrupting British naval activity on the western waters of Lake Ontario.

Despite the British buildup, Winfield Scott asked permission from General Brown to take his brigade and advance around the lake toward York. Brown declined to sanction what may well have become a reckless romp and instead ordered Scott and all of his command to withdraw from Queenston Heights and retreat south past the falls to a position once again along the Chippawa. But this, too, left Brown uneasy. There were reports that Drummond was dispatching a sizable force to the American side of the Niagara, perhaps in an attempt to outflank him or to plunder what was left of Buffalo. Not sure of what was going on, Brown had scarcely reached the Chippawa when he ordered Scott to take his brigade and once again reconnoiter north toward

Queenston Heights. Perhaps recognizing his subordinate's greatest strengths as well as his limitations, Brown admonished Scott "to report if the enemy appeared, and to call for assistance if that were necessary."[7]

Scott's brigade—Scott with the advance guard as usual—marched north from the Chippawa along the Portage Road the few miles to Niagara Falls. Reaching Willson's Tavern, which overlooked the Horseshoe Falls at Table Rock, in the late afternoon of July 25, Scott discovered British officers fleeing in much the same haste that he had fled his uneaten breakfast at Mrs. Street's farmhouse a few weeks before. Several of the officers tossed him courteous salutes as they galloped away, and Scott entered the tavern to quiz its owner, the widow Willson.

Yes, Deborah Willson offered graciously, the British were formed on Lundy's Lane just around the bend with more than a thousand men and two cannon. But Scott was skeptical. He still assumed that the main British force was on the eastern bank of the Niagara and that he could easily push aside whatever troops stood between him and Queenston Heights. Poor Winfield Scott. He wasn't having much luck with the ladies. Mrs. Street had conned him and now when Mrs. Willson told him the truth, he didn't believe her! So Scott's brigade marched boldly toward Lundy's Lane.[8]

Almost opposite Niagara Falls, the little dirt track of Lundy's Lane ran west from the Portage Road atop a low rise crowned by a church used by several denominations. Below the church, a large open field extended south to a chestnut forest. The field offered little protection to anyone moving across it. Much to his chagrin, Winfield Scott found this out the hard way. As his brigade marched into the field, they came into "full view, and in easy range of a line of battle drawn up in Lundy's Lane, more extensive than that defeated at Chippawa." The brigade was subjected to a withering fire from the British line, and it was immediately obvious to Scott that he had blundered into a major British force. Retreat, however, did not occur to him, and he hastily sent word to Brown to bring up Ripley's and Porter's brigades. Had he retreated, the entire American army may have followed suit. But as it was, "by standing fast, the salutary impression was made upon the enemy that the whole American reserve was at hand and would soon assault his flanks." Indeed, Generals Drummond and Riall thought so. Drummond ordered up Colonel Hercules Scott's 103rd Foot Regiment and called for his own reserves to tighten his line.[9]

One thing was certain. Scott could not keep his brigade in this field forever. Drummond's artillery was positioned in the cemetery below the church, and even through the thick smoke of gunpowder that quickly enveloped the field, the regimental standards of Scott's units made for inviting targets. Determined to do something, Scott ordered an advance, thought better of it, and then dispatched Major Thomas S. Jesup and his Twenty-fifth Infantry to attack the junction of Lundy's Lane and the Portage Road and attempt to turn the British left flank. Just as he did at Chippawa, Jesup gamely moved forward.

By now, even in the midsummer twilight, it was getting dark. One of the first officers to fall wounded on the British side was General Riall, who would later lose his right arm. His aides guided him down Lundy's Lane toward the Portage Road en route to Queenston and shouted out at a cluster of troops at the junction to "Make way for the General!" The group cleared the roadway with acknowledgments of "Yes, sir, yes, sir" but quickly formed again to surround the general. "What is the meaning of this?" Riall demanded. "You are our prisoner, sir," replied Captain Daniel Ketchum of Jesup's Twenty-fifth Infantry.[10]

Meanwhile, the remainder of Scott's brigade was in sad shape, but before it could be driven from the field, General Brown arrived on the scene with reinforcements. Still anticipating an attack from the phantom British forces on the eastern bank of the Niagara, Brown had left Porter's Third Brigade in reserve and advanced with only Ripley's Second Brigade. But it might be enough to turn the tide. Brown ordered the undermanned First

Infantry Regiment with only 150 men to feint toward the center of the British line below the church. Then Brown rode to find Lieutenant Colonel James Miller of the Twenty first Infantry, one of the most battle-hardened units in the American army. Miller had cut his teeth at Tippecanoe, fought near Detroit, and followed Wilkinson's folly down the St. Lawrence. Within the hour, Miller would lead the Twenty-first into legend.

Recognizing that seizing the British artillery in the churchyard was the key to the battleground, General Brown rode up beside Miller and directed him to storm the guns and take the position. Miller was physically imposing and single-minded of purpose. He squinted for a moment at the British battery and then turned to eye Brown. His reply of "I'll try, sir!" was to become the motto of the regiment.

But even before Miller's regiment could start up the hill, Brown was startled to see the First Infantry advancing on the same guns. Somehow, the First had misunderstood its orders to conduct only a feint. Before its commander, Lieutenant Colonel Robert C. Nicholas, rethought the matter and ordered his men to retreat, the First's efforts had ended up doing exactly as Brown had hoped and distracted the British while Miller's regiment moved smartly up the slopes with its bayonets at the ready. One British observer later reported that Miller's men "charged to the very muzzles of our cannon and actually bayoneted the artillerymen who were at their guns."[11]

Not only did the Twenty first capture Drummond's artillery, but it surprised Morrison's Eighty-ninth Foot, which had been in a line just north of the crest of the hill. Sensing a general rout, Drummond ordered Morrison to pivot his line so as to catch the lost artillery position and the advancing Americans in a cross fire. This was done rather smartly, but instead of faltering, the Twenty-first returned volley after volley of musket fire. No wonder that Miller himself later described the scene as "one of the most desperately fought actions ever experienced in America."[12]

The Eighty-ninth Foot finally retreated from the hilltop, but were they finished? General Brown rode up to congratulate Colonel Miller and survey the situation. Brown was certain that his men had won a resounding victory, but Ripley was equally certain that the British were preparing a counterattack. Determined to see for themselves, Brown, Ripley, and their aides rode west along Lundy's Lane in front of the American line. Now it was Brown's turn almost to be taken prisoner because of a case of mistaken identity in the darkness. A regimental line appeared through the murky night, but was it friend or foe? Brown's aide, nineteen-year-old Captain Ambrose Spencer, spurred his horse forward to investigate. "What regiment is that?" Spencer cried out. "The Royal Scots, sir," came the prompt answer. "Stand you fast, Scotch Royals," replied Spencer, as he wheeled his horse and scurried Brown's party back to the American lines.[13]

So the British were indeed counterattacking. Drummond really had no choice. His army's artillery had been captured, and without it he might as well retreat to Fort George. Throughout the darkness as the hours wound toward midnight both sides continued to attack and counterattack. Winfield Scott, though wounded, was back in the fray leading the remnants of his brigade west along Lundy's Lane. In the process, his troops were fired upon by both sides. Almost at the same time, General Brown was wounded in the thigh and his daring aide, Ambrose Spencer, killed. There followed a final counterattack by the British that resulted in hand-to-hand fighting over and around the gun carriages. Scott was hit by a ball that shattered his shoulder, and unsure that he was even alive, his men carried him to the rear. Among those to fall at the point of this farthest advance was Captain Abraham Hull of the Ninth Infantry, the son of General William Hull, who had surrendered Detroit so many months before.

As Scott was making his final attack, General Brown received a second wound. Barely able to stay in his saddle, the general tried to find Scott to turn command over to him. Told that Scott had been carried to the rear, Brown sent his surviving aide, Captain

Loring Austin, to find General Ripley and order him to assume command. At this point near midnight, the Americans had driven the British off Lundy's Lane and retained control of their artillery in the churchyard.

What happened next has long been a matter of great debate on both sides of the field. As Brown rode slowly toward the rear, a number of his officers counseled a general withdrawal back to Chippawa to regroup. Others advocated keeping possession of the hard-won position. One who chanced to speak with Brown as he rode away from the carnage was Major Jacob Hindman, his chief of artillery. Hindman was hurrying forward with additional ammunition for both his own artillery pieces and the captured British ones. Whatever Brown said—and whether it was said conversationally or as a direct order—Hindman interpreted the sentiments as an order to withdraw and passed on such to Ripley.

As the ranking American officer, Ripley now had his own cluster of officers who advocated withdrawal or consolidation of a position upon the hilltop. In the end, Ripley decided that it was safer to follow Brown's purported order to withdraw—no matter how disputed—rather than take the responsibility for exposing his badly mangled brigades to another British attack. The end result was that the Americans withdrew from the field in the wee hours of July 26, inexplicably leaving the British artillery in place. Imagine General Drummond's surprise a few hours later when the morning's light not only revealed no sign of the Americans but also showed his precious artillery left unattended. Once more in possession of the field of battle, Drummond quickly reclaimed his artillery and declared victory.[14]

As always in such nineteenth-century encounters, exact casualty totals were difficult to determine. British losses were reported as 84 killed, 559 wounded, 193 missing, and 42 taken prisoner, a casualty rate of roughly 25 percent of the force engaged. Morrison's Eighty-ninth Foot had been particularly hard hit and lost more than 60 percent of its effective strength. On the American side, losses were given as 173 killed, 571 wounded, 38 missing, and 79 captured, more than 30 percent of those engaged. Scott's First Brigade and Miller's Twenty-first Infantry bore the brunt of these numbers. While such totals may seem slight in the face of the slaughter that was to occur on American fields two generations hence, the high percentage of casualties for those engaged made what came to be called the Battle of Lundy's Lane one of the bloodiest of the war. "We boast of a 'Great Victory,'" wrote British Colonel Hercules Scott of the 103rd Foot, "but in my opinion it was nearly equal on both sides." One thing was clear, the ragtag American army was learning how to stand and fight.[15]

But now the Americans were retreating. However confusing his orders or intent in the early morning of July 26, General Brown now determined that he must withdraw to the safety of Fort Erie. On the British side, General Drummond was content for the moment to let him do so. Drummond, too, had been wounded, and that fact only served to remind him how badly his army had been mauled, even if it had managed to retain its artillery. A few days later, as Drummond cautiously advanced south across the Chippawa battlefield toward Fort Erie, Ripley pleaded with a convalescing Brown in Buffalo that the Americans should abandon the fort and the entire Canadian side of the Niagara. No, said Brown, Ripley was to stand firm and hold the fort at all costs. This time Ripley made certain that he got his orders from Brown in writing.[16]

The Americans had in fact strengthened Fort Erie during the summer of 1814, and despite Ripley's concerns in holding it, Drummond did not fancy a frontal assault against it. Instead he hoped to persuade the Americans to withdraw completely from the Canadian side of the river by making a raid against the American side. On August 3 Drummond sent Lieutenant Colonel John Tucker with a force of six hundred men across the river to disrupt supply depots at Black Rock and Buffalo. They met with stout resistance from three hundred Americans at Conjocta Creek and were forced to withdraw without accomplishing their objective. So Drummond was back to square one and forced to order that which he abhorred.

In the early morning of August 15, 1814, under the cover of a heavy downpour, three columns of British troops, including Colonel Hercules Scott's 103rd Foot, moved forward against Fort Erie with bayonets fixed. The two northern columns succeeded in breaching the bastion on the north wall despite heavy hand-to-hand fighting. Colonel Scott was among those to fall—dead almost instantly from a bullet to the head. The British managed to swing one American cannon around to fire into the fort. In the process, however, sparks from its muzzle blast fell through the wooden flooring and ignited a stockpile of gunpowder. The result was a tremendous explosion that shook the ground for miles and sent bodies flying through the night. The bastion and a goodly portion of Drummond's attacking troops were blown sky high. When the smoke cleared the following morning, the British had suffered more than nine hundred casualties out of an attacking force of about twenty-five hundred—more than at Lundy's Lane—and the Americans remained in possession of Fort Erie.[17]

Throughout a dismal, rainy September, Drummond pondered what to do next. His supply lines were finally being threatened by Isaac Chauncey's belated appearance with his fleet in the western waters of Lake Ontario. Reinforcements of New York militia were seen rowing across the Niagara from Buffalo almost daily. If the Americans were that determined to hold Fort Erie, so be it. Drummond decided to withdraw and give up the siege, only to have the Americans storm out of the fort and attack his positions just as he was removing his artillery. It was another bloody affair that accomplished little despite one American officer's account that it was "the most spendid achievement" of the campaign. For his part, General Brown was heartened by the courage and discipline displayed by the much-maligned New York militia. "The Militia of New York," the general boasted, "have redeemed their character."[18]

But that was about all. The Americans marched back to Fort

Erie. The British finished packing up their cannon and hauled them back to Fort George. The bloody travail of the war's third Niagara campaign was over, having failed to accomplish any more on either side than the previous two. Regardless of the recurring arguments over which side won the Battle of Lundy's Lane, Drummond's stand that day had blunted Brown's drive down the Niagara and once and for all ended any hope that the Americans would conquer Upper Canada. On November 5, 1814, Major General George Izard, who had taken over from General Brown, ordered the Canadian side of the Niagara evacuated and Fort Erie blown up. Since then, the only thunder along the Niagara has been peaceful.

LAKE CHAMPLAIN

The water corridor between New York and Montreal via Lake Champlain and the Hudson and Richelieu rivers was long the scene of conflict between those determined to control half a continent. During the French and Indian War, French colonists on the St. Lawrence and their British counterparts along the Hudson fought up and down it. In the opening days of the American Revolution, Richard Montgomery and Benedict Arnold led colonial troops—including young officers James Wilkinson and Aaron Burr—from the Lake Champlain country to the very gates of Quebec. Two years later, in 1777, British general "Gentleman Johnny" Burgoyne marched south from Canada along the lake route, determined to reach New York City and sever the New England limbs of the Thirteen Colonies. That Burgoyne ended up surrendering his army at Saratoga seems not to have been lost upon the British as they began a similar thrust south during the summer of 1814.

The man charged with the exercise was General Sir George Prevost, since 1811 governor-general of Canada and commander in chief of His Majesty's forces in Canada, the Atlantic Colonies, and Bermuda. During the first two years of war, Prevost's task had been to hold Canada together administratively and defend it territorially, all the while juggling the incessant pleas of various theater commanders for ever more troops and matériel—both of which were always in short supply. That Prevost was largely successful in this is evidenced by the fact that save for American dominance on Lake Erie, the Canadian–American boundary was essentially the same as it had been in June 1812. Now with the British lion's patience run out and an abundance of veterans available from the Napoleonic campaigns, Prevost was finally to have both the troops and the supplies he needed to conduct offensive operations.

On June 3, 1814, the Earl of Bathurst, secretary of state for war and the colonies, and as such the principal civilian architect of British policy in North America, sent General Prevost his marching orders for the summer. Embarking from Bordeaux, France, bound for Canada were "twelve of the most effective Regiments of the Army [that had served] under the Duke of Wellington together with three Companies of Artillery on the same service." When joined by other units en route from Europe, this force would approach ten thousand veteran infantry—by far the largest and most battle-tested British force ever to assemble in North America. With this force under Prevost's command, Bathurst clearly expected great things.

"His Majesty's Government conceive that the Canadas will not only be protected for the time against any attack which the Enemy may have the means of making," instructed Bathurst, "but it will enable you to commence offensive operations on the Enemy's Frontier before the close of this campaign." Protecting Canada meant, Bathurst went on to say, finally destroying the pesky American base at Sackets Harbor and regaining naval superiority on Lake Erie and Lake Champlain. Contemplated offensive operations included the recapture of Detroit and the Michigan country, the "retention of Fort Niagara and so much of the adjacent Territory as may be deemed necessary," and, if possible, the establishment of "any advance position on that part of our frontier which extends towards Lake Champlain."

Showing that he for one remembered the embarrassment of Saratoga, Bathurst added that Prevost should take care in executing the latter so as not to "expose His Majesty's Forces to being cut off by too extended a line of advance."[1]

If Bathurst contemplated an eventual march to New York City and the severing of New England—the same grand strategy that had failed a generation before there was no hint of such on the face of his June 3 orders. Perhaps he recognized that the first step was to get Prevost moving in the right direction. Prevost acknowledged Bathurst's orders on July 12 and wrote that they would be obeyed as soon as his full force had assembled, but that "defensive measures only will be practicable, until the complete command of Lakes Ontario and Champlain shall be obtained, which cannot be expected, before September."[2]

That was hardly the response the Earl of Bathurst had expected, considering that the might of the British Empire was now sailing up the St. Lawrence. "If you shall allow the present campaign to close without having undertaken offensive measures," Bathurst replied icily, "you will very seriously disappoint the expectations of the Prince Regent and the country."[3]

Doubtless Prevost had already surmised as much as he assembled his army south of Montreal. Despite Bathurst's promise of provisions for ten thousand men for six months, Prevost now had more than twenty-nine thousand troops spread throughout Canada. To feed them, he found assistance from an unlikely source. "Two-thirds of the army," Sir George reported to Bathurst on August 27, "are supplied with beef by American contractors, principally of Vermont and New York."

It was true. Yankee farmers were putting British gold ahead of American patriotism and sending New England beef and produce flowing north to Canada. Among those on the American side enraged by this was Major General George Izard, who received a letter from one Vermonter bemoaning that "droves of cattle are continually passing from the northern parts of this state into Canada for the British."

Izard forwarded the letter to the War Department and added his own commentary: "This confirms a fact not only disgraceful to our countrymen but seriously detrimental to the public interest. From the St. Lawrence to the ocean an open disregard prevails for the laws prohibiting intercourse with the enemy. . . . On the eastern side of Lake Champlain the high roads are insufficient for the cattle pouring into Canada. Like herds of buffaloes they press through the forests, making paths for themselves. Were it not for these supplies, the British forces in Canada would soon be suffering from famine." Small wonder that Prevost acknowledged that "Vermont has shown a disinclination to the war, and as it is sending in specie and provisions, I will confine offensive operations to the west [New York] side of Lake Champlain."[4]

As Sir George's massive army prepared to do just that, the bulk of the American troops opposing him—some four thousand regulars under General Izard's command—marched off for Sackets Harbor under belated orders from the War Department that presumed Prevost would strike there first. To no avail, Izard objected strenuously. Brigadier General Alexander Macomb was left to defend the western shores of Lake Champlain with a mismatched assemblage of three thousand regulars and militia, only half of whom were fit for duty. As he left for Sackets Harbor, Izard warned that all of New York north of recently erected works on the Saranac River at Plattsburgh would be in the possession of the enemy "in less than three days after my departure."[5]

Actually, it took Prevost a few days more than that. Crossing the border on August 31, 1814, Prevost reached Chazy, twenty-five miles north of Plattsburgh, on September 4. General Macomb's men felled trees, broke bridges, and engaged in skirmishing to try to stem the British advance, but these tactics did little to slow the onslaught of Wellington's veterans. "They never deployed [spread out] in their whole march," Macomb reported, "always pressing on in column." By the evening of September 6, the British were in Plattsburgh, and Macomb retreated south across the Saranac River to fortifications that overlooked Plattsburgh Bay. Now the British paused and spent four days preparing to assault the American position—tenuously held though it was—while Prevost looked out across the waters of Lake Champlain for some assistance from the Royal Navy.[6]

• • •

The balance of naval power on Lake Champlain had rested with the United States until June 1813 when the overzealous Lieutenant Sidney Smith caused the sloops *Growler* and *Eagle* to be lost to the British. This left the Americans to retreat to the upper end of the lake and frantically set about building new ships. (While Lake Champlain runs north and south, it drains north to the St. Lawrence; thus, the southern area is in fact the *upper* lake.) The British did the same, but unlike Chauncey and Yeo, who had engaged in a shipbuilding race on Lake Ontario, the officers at Lake Champlain were determined to fight once ships were built.

The American commander on the lake, Master Commandant Thomas Macdonough, was certainly no stranger to its waters. Born in Delaware in 1783, Macdonough joined the navy as a mid-shipman at sixteen. He saw service in the Mediterranean under Stephen Decatur and participated in the burning of the captured *Philadelphia*. Tall and slender, he became a staunch, God-fearing Episcopalian. By 1807, he was first lieutenant of the *Wasp* and enforcing Jefferson's embargo along the Atlantic coast. Afterward, Macdonough took a lengthy leave to captain a merchantman to Calcutta and back before once more applying for active duty at the outbreak of war. He was initially posted to the *Constellation* as its first lieutenant. But the frigate he had once served on as a young midshipman proved unfit for sea, and he was quickly given command of a small fleet on Lake Champlain, arriving there early in October 1812.

After *Growler* and *Eagle* were lost the following June, Macdonough was forced to play cat-and-mouse with the British squadron commanded by Royal Navy Lieutenant Daniel Pring. Macdonough was definitely the mouse as Pring ranged about the lake and even provided transport for British troops to raid Plattsburgh and Burlington. Macdonough confined his forays to the vicinity of the shipyard at Otter Creek near Vergennes, Vermont. By fall both Pring and Macdonough were focused on shipbuilding.[7]

The centerpiece of Macdonough's new fleet was the ship

Saratoga, launched at Otter Creek, on April 11, 1814. Her name was no small coincidence and was meant as a proud and defiant reminder of America's greatest victory to date save Yorktown. Larger than Perry's *Lawrence* or *Niagara*, *Saratoga* at seven hundred tons was armed with eight long 24-pounders and six 42-pound and twelve 32-pound carronades, capable of throwing a 414-pound broadside. Manned by a crew of 210, she was not exactly "Old Ironsides," but a formidable force on Lake Champlain nonetheless.

To accompany *Saratoga*, Macdonough could rely on the hastily completed brig *Eagle* under the command of Captain Robert Henley. Comparable in size to the *Lawrence* or *Niagara*, *Eagle* was armed with eight long 18-pounders and twelve 32-pound carronades. The schooner *Ticonderoga* was originally a small steamship—still a novelty, as Fulton's *Clermont* had debuted only in 1807—but constant problems with the steam machinery forced her to rely on sails and deprived her of the distinction of being America's first steam-powered warship. At 350 tons, *Ticonderoga* carried eight long 12-pounders, four long 18-pounders, and five 32-pound carronades. Finally, among his capital ships Macdonough counted the small sloop *Preble*, mounting seven long 9-pounders. These four vessels were accompanied by an assortment of ten gunboats, each with one or two long guns. That totaled fourteen vessels throwing 1,194 pounds of broadside, almost two-thirds of which was from short-range carronades. By May, even as the *Eagle* was still being completed and despite Pring's efforts to attack his base at Otter Creek, Macdonough was sailing this flotilla off Plattsburgh and was once again in control of the upper lake.[8]

Pring had a superweapon of his own under construction, but he was not to command her. Despite his knowledge of the lake, Pring had managed to run afoul of Sir James Yeo, still the Royal Navy's commander in chief on the Great Lakes and surrounding waters. Perhaps because he thought Pring a little too aggressive, Yeo replaced him almost on the eve of battle—first with Captain Peter Fisher, who lasted but six weeks in Yeo's good graces, and then with Captain George Downie. Two years older than Macdonough, thirty-three-year-old Downie had had a similar career and shared an Irish heritage. Born in the county of Ross, Ireland, Downie entered the Royal Navy as a midshipman and served on a number of vessels, particularly in the West Indies. By 1804 he was a lieutenant aboard the frigate HMS *Sea Horse* and in due course worked his way up to post captain, the rank he held when the Admiralty dispatched him to Montreal in command of a squadron of transports carrying components for Yeo's shipbuilding efforts on Lake Ontario. Yeo sent him to Lake Champlain at the eleventh hour.[9]

Downie's new flagship—and the match for the *Saratoga*—was the twelve-hundred-ton, 160-foot *Confiance* (Confidence), named by Sir James Yeo not out of any idle boast—Yeo certainly had no room to make any—but rather out of mere sentiment as the namesake of his first command. Planned for a crew of 325, she carried thirty long twenty-four-pounders, six thirty-two-pound carronades, and a long twenty-four on a pivot. Her broadside weight bested *Saratoga* by at least sixty-six pounds and perhaps as many as ninety-six. (There is some evidence that her carronades were actually forty-two-pounders.) But what really provided an advantage was an onboard furnace for heating shot. *Confiance* had been on the water only a week when Downie stepped aboard her and assumed command of the Lake Champlain squadron on September 2, 1814. Carpenters were still at work on his flagship, and she was as new and green to the lake as he was.

Joining *Confiance* were the brig *Linnet* and two sloops. The eighty-five-foot *Linnet* remained under Lieutenant Pring's command and was about the size of the *Ticonderoga*, mounting sixteen long twelve-pounders. If the sloops *Chub* and *Finch* looked familiar, they should have. These were the former American sloops *Eagle* and *Growler* lost the year before by Lieutenant Smith. *Chub* carried ten eighteen-pound carronades and one long six-pounder; and *Finch* six eighteen-pound carronnades, four long six-pounders, and one short eighteen-pounder. The British also had an assortment of twelve gunboats, the larger ones of which carried both a long gun and one carronade.

In all, this amounted to sixteen British vessels throwing a broadside of 1,192 pounds—at first blush almost exactly equal to that of the Americans. But upon closer inspection, the allotment of long guns versus carronades was almost the reverse of the Americans, the British fleet having 55 percent of its armament in long guns. Again, Theodore Roosevelt in *The Naval War of 1812* painstakingly did the math and the analysis, but the bottom line was simple: out on the lake at long range, Downie held an advantage. At close range, where his carronades could pound away, the advantage belonged to Macdonough.[10]

The biggest difference between the two fleets, however, was that time was on Macdonough's side. He held the southern lake and could afford to wait. Let the British come. It was Downie who was under pressure from General Prevost to do something. Accordingly, Macdonough sought out the deep waters of Plattsburgh Bay. Later, there would be great debate over how important Macomb's artillery on the western shores of the bay was to supporting Macdonough's anchorage and vice versa, but the bay itself was a natural defensive position. Roughly two miles wide, it opened to the south with the main shore and mouth of the Saranac River to the west and Cumberland Head dividing it from the main lake on the east. This meant that any north wind favorable to the British for sailing up the lake (to the south) would quickly become a headwind as their ships rounded the head and entered the bay. Crab Island and some extensive shoals extended from the mainland at the mouth of the bay and further restricted the movements of any ships entering it.

Macdonough took advantage of this geography to anchor his fleet just inside the mouth of the bay in a line extending roughly southwest to northeast between Crab Island and Cumberland Head. This meant that any enemy entering the bay would likely have to do so bows on—sailing straight into his broadsides. The *Eagle* lay on the northern end of the line, close enough to Cumberland Head that any attempt to run between her and the shore would place the opponent under heavy fire. *Saratoga* came next in line, but as his heavy weapon, Macdonough arranged a system of anchors and winches that permitted the ship to be pivoted to face any direction that future circumstances should warrant. *Ticonderoga* was anchored south of *Saratoga*, followed by *Preble*, whose task it was to guard the shallow passage between the main fleet and the shoals of Crab Island and prevent any flanking maneuver from that direction.[11]

Just why Sir George Prevost was so determined to wait for the Royal Navy and have Downie sail into this hornet's nest is debatable. Certainly, his army of more than ten thousand veterans should have been easily able to steamroller over Macomb's troops no matter how effective their breastworks above the Saranac may have been. That is exactly what Prevost's brigade commanders, major generals Frederick Robinson, Thomas Brisbane, and Manley Power, expected him to order. But whereas they had made their marks on the battlefields of Spain under the Iron Duke, Prevost had made his in administrative and defensive operations in North America. To take orders from someone who had only colonial experience, and was younger than any of them at that, galled to say the least. But at least let there be orders. "It appears to me," Robinson later confessed to his journal, "that the army moved against Plattsburgh without any regularly digested plan by Sir George Prevost."[12]

Actually, Sir George considered ordering Robinson's brigade to storm the American position upon arriving in Plattsburgh, but hesitated when his intelligence failed to report the whereabouts of fords across the Saranac or the distance of the American fortifications beyond its bank. By the time this information became available, Prevost had determined to wait for Downie. "It is of the highest importance that the ships, vessels, and gunboats, under your command," Prevost wrote to Downie shortly after Downie's arrival on the lake, "should combine a co-operation with the division of the army under my command. I only wait for your arrival to proceed against General Macomb's last position on the south bank of the Saranac."[13]

Downie replied that he was willing to do so, but that *Confiance* was far from ready for action. Carpenters were still at work, and he was doing his best to scrape together a crew. Hastily assembled, it was to be like that of the *Chesapeake* when Lawrence hurriedly sailed out to engage *Shannon*, unknown to the ship and largely unknown to one another. "Until she is ready," Downie told Sir George, "it is my duty not to hazard the squadron before an enemy who will be superior in force."[14]

Why Downie thought Macdonough's force superior is open to question. Perhaps it was merely his way of buying a few extra days. Certainly the *Confiance* needed every hour that could be bought, and in her rookie state she may well have been inferior to Macdonough's seasoned hands aboard *Saratoga. Confiance* was still being rigged as she was towed up the lake to join the remainder of Downie's squadron off Chazy. "It scarcely needs the habit of a naval seaman to recognize," recounted Admiral Alfred Thayer Mahan long after the fact, "that even three or four days' grace for preparation would immensely increase efficiency."

But Downie was not to have those days. Now it was Sir George's turn to adopt an icy tone. The next day, September 9, Prevost replied: "In consequence of your communication of yesterday I have postponed action until your squadron is prepared to co-operate. I need not dwell with you on the evils resulting to both services from delay." Prevost was senior in rank and years, not to mention being governor-general of Canada. What was poor Downie to do?

Beaten down, the young captain prepared to sail at midnight on the evening of the ninth, but strong headwinds prevented him from doing so. When he learned of this most recent delay, Sir George's words carried even more bite than the wind. Noting that his troops were ready to storm Macomb's position at the same moment as Downie commenced his attack, Prevost chastised: "I ascribe the disappointment I have experienced to the unfortunate change of wind and shall rejoice to learn that my reasonable expectations have been frustrated by no other cause."

"This letter does not deserve an answer," remarked Downie hotly to Pring, "but I will convince him that the naval force will not be backward in their share of the attack." Downie arranged to scale his guns—fire powder cartridges without shot to clean out the bores—to signal his approach so that Sir George would know when to begin his coordinated attack. He hoped this would cause the American ships to flee their protected anchorage or at the very least cause them confusion. To Pring, however, Downie admitted that he would much rather fight on the open waters of the lake.[15]

In the wee hours of September 11, 1814, with the wind from the northeast, Downie's little squadron swept up the lake. At five o'clock, upwind and at a distance of some six or seven miles from Prevost's position, Downie scaled his guns. There! Let Sir George know that the Royal Navy was en route and prepared to do its fair share. Two and one-half hours later off Cumberland Head, the British squadron hove to so that the smaller gunboats could catch up to the bigger vessels. Downie took advantage of the pause to be rowed in his gig around the tip of Cumberland Head. This allowed him to peer into Plattsburgh Bay and personally ascertain the American position.

The order of the American line from north to south was just as Macdonough had planned: *Eagle*, *Saratoga*, *Ticonderoga*, and *Preble*. Lest anyone forget after two years of war what all the fighting was about, Macdonough ran a signal up the mast of his flagship that proclaimed, "Impressed seamen call on every man to do his duty."[16]

Downie's plan was to sweep around Cumberland Head with *Finch* leading *Confiance*, *Linnet*, and *Chub*. When the line tacked to starboard and turned into the wind to sail into the bay, this would put *Linnet* and *Chub* in the van to attack *Eagle*. *Confiance* would move out in front of the line, assist with a broadside against *Eagle*, and then anchor across the *Saratoga*'s bow and rake her. With the three strongest British ships arrayed against the two strongest Americans, *Finch* and the gunboats would be left to harry *Ticonderoga* and *Preble* and keep them out of the main action. Meanwhile, Downie fully expected Prevost's men to have seized the heights above the Saranac

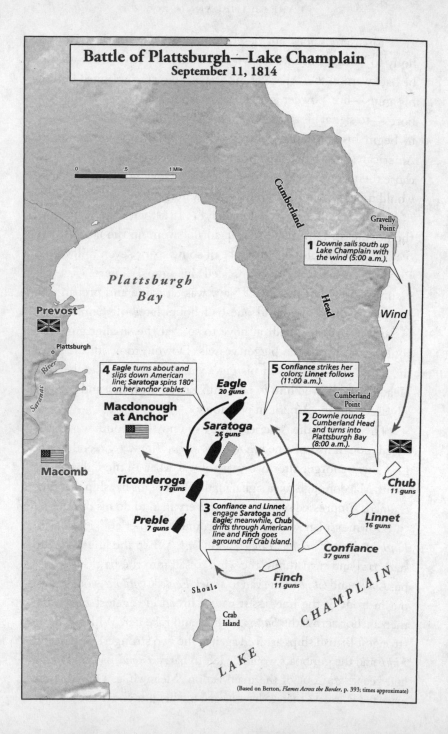

Battle of Plattsburgh—Lake Champlain
September 11, 1814

0   .5   1 Mile

Cumberland

Gravelly
Point

1 Downie sails south up
Lake Champlain with
the wind (5:00 a.m.).

*Plattsburgh
Bay*

Head

Wind

Prevost

Plattsburgh

Cumberland
Point

4 *Eagle* turns about and
slips down American
line; *Saratoga* spins 180°
on her anchor cables.

5 *Confiance* strikes her
colors; *Linnet* follows
(11:00 a.m.).

Eagle
20 guns

**Macdonough
at Anchor**

2 Downie rounds
Cumberland Head
and turns into
Plattsburgh Bay
(8:00 a.m.).

Saratoga
26 guns

River

Saranac

Macomb

Ticonderoga
17 guns

Chub
11 guns

3 *Confiance* and *Linnet*
engage *Saratoga* and
Eagle; meanwhile, *Chub*
drifts through American
line and *Finch* goes
aground off Crab Island.

Linnet
16 guns

Preble
7 guns

Confiance
37 guns

Shoals

Finch
11 guns

Crab
Island

LAKE CHAMPLAIN

LAKE

(Based on Berton, *Flames Across the Border*, p. 393; times approximate)

River and turned the captured American batteries onto the rear of the American fleet. The American vessels would be either trapped or forced to run for the open lake, where Downie's long guns would show their advantage.

Confidence! It might work after all. But as *Confiance* rounded Cumberland Head and headed upwind to lead the attack, there was no evidence of Sir George's army rushing to attack the heights. Downie quickly became painfully aware that he was alone. Alone, but committed. Whatever thoughts he now had for Sir George would soon go with him to his grave.

The British line stood into Plattsburgh Bay, *Linnet* leading. The first broadside from her long twelves fell short of *Saratoga* except for one round. That one shot hit a hen coop on the *Saratoga*'s deck and blew it apart, releasing the gamecock that was inside. Instead of cowing, the cock flapped his wings and crowed loudly. The *Saratoga*'s crew cheered just as loudly, and Macdonough fired the first gun in reply.

Confiance's first broadside is said to have struck down a fifth of the *Saratoga*'s crew, evidence that had a duel occurred between these two ships in open waters, *Confiance*'s heavier broadside weight might have told the tale. But in the narrow confines of Plattsburgh Bay, the wind quickly died and Downie's flagship was forced to anchor some five hundred yards from the American line. Pring's *Linnet* reached the *Eagle* and closed with her, but *Chub*'s support was not forthcoming. *Chub* suffered extensive damage to her sails and rigging and ended up drifting through the American line before she could anchor. *Finch* had trouble keeping into the wind and failed to close with *Ticonderoga*. She eventually went aground on Crab Island. British gunboats scurried about and forced *Preble* from her anchorage at the end of the American line. Thus, the deciding contest quickly shaped up to be between *Saratoga* and *Confiance*, and *Eagle* and *Linnet*.

The fighting was fast and furious. Twice the *Saratoga* was set afire by hot shot heated in the *Confiance*'s furnace. Although out-weighted, *Saratoga* answered *Confiance* broadside for broadside.

One American shot struck the muzzle of a British cannon and knocked it off its carriage. It fell against Captain Downie, pinning him dead and flattening his watch in the process. *Eagle* soon became so pummeled that Captain Henley cut her anchor cable and caught enough wind in her topsails to drift south around *Saratoga*, turning about as he did so. Once back in the line, Henley reanchored *Eagle* and brought her undamaged side to bear on the enemy. This enabled *Eagle* to pour fire into *Confiance*, but it also freed up *Linnet* to hammer *Saratoga* from the north.

Now Macdonough did a similar maneuver. Without getting under way, he spun his ship 180 degrees using winches and the anchor chains and brought her uninjured broadside to bear on *Confiance*. Downie's executive officer, Lieutenant Henry Robertson, who assumed command of the *Confiance* upon Downie's death, attempted a similar maneuver. *Confiance*, however, had dropped anchor quickly upon entering the fray, and without the careful positioning and rigging Macdonough had done aboard *Saratoga*. Robertson was able to execute only half of the turn, leaving his ship at right angles to a deady raking fire. "The ship's company," reported Robertson afterward, "declared they would stand no longer to their quarters, nor could the officers with their utmost exertions rally them."

Water was pouring into *Confiance*. Her guns were run in on the port side to throw some weight to starboard and bring the cannonball holes on her port side out of the water. Much of *Saratoga*'s rigging and several of her masts were shot away, but she was clearly the victor. Lieutenant Robertson had no choice but to strike his colors at 11:00 A.M., more than two hours after the engagement began. Pring struck *Linnet*'s colors fifteen minutes later, and the Battle of Lake Champlain was over. There was still no sign of General Prevost's troops on the ramparts above the bay.

According to Macdonough, *Saratoga* had 55 round shot in her hull, *Confiance* 105. Casualties were particularly difficult to calculate, in part because of the variety of last-minute recruits on both sides. Perhaps the most reliable estimate is fifty-two Americans killed and fifty-eight wounded, and fifty-seven British killed and seventy-two wounded. Seventeen British officers and upwards of three hundred seamen were also taken prisoner. Captain Macdonough graciously returned the proffered swords of the surviving British officers, and Lieutenant Pring was glowing in his praise of the Americans' prompt and cordial treatment of the British wounded.

Placing his trust first and foremost where it had always been, Macdonough reported that "the Almighty has been pleased to grant us a signal victory on Lake Champlain in the capture of one frigate, one brig, and two sloops of war of the enemy." Echoing that sentiment, no less a naval authority and historian than Admiral Alfred Thayer Mahan wrote almost a century later that "the battle of Lake Champlain, more nearly than any other incident of the War of 1812, merits the epithet 'decisive.'" But where was Sir George Prevost?[17]

Major General Frederick Robinson's brigade and that of Major General Manley Power had been waiting since an hour before dawn to cross the Saranac and attack Macomb's left flank, planning to drive him from his position. Robinson himself had heard Downie scale his guns while riding to Prevost's headquarters to receive his final orders. But Prevost ordered that the attack not begin until 10:00 A.M. When Robinson finally led his troops forward at that late hour, there was still confusion as to which road to take to ford the Saranac. An hour passed before Robinson neared the ford, and by then, he recorded, "we heard three cheers from the Plattsburgh side." Sending an aide to ascertain which side was cheering, Robinson soon realized how dreadfully late Prevost's orders had been.[18]

Prevost realized it, too. With thoughts of Saratoga no doubt echoing in his brain, he finally moved with alacrity—but in the direction not even General Macomb had expected. Dispatching a hasty order to Robinson, Prevost noted the defeat of Downie's fleet and ordered him to "immediately return with the troops under your command." If Robinson was surprised, Macomb was even more so. By nightfall, the largest British army ever to tread

American soil had burned surplus stores and munitions and started back to Canada. Macomb hardly had time to blink, and they were gone. He reported to the War Department on the valiant defense his men had made, but it was really Sir George's order to retreat that had won him the day.[19]

Unlike the Battle of Lake Erie where the victors quarreled over the roles of Perry and Elliott, on Lake Champlain it was the vanquished who refought the battle with a vengeance. Lieutenant Pring, though he had done all that he could to support Downie and Prevost's rushed strategy, was hauled before the requisite court-martial. Pring was exonerated only when the navy decided to blame the army for its loss. Sir George Prevost was summoned home to England to face his own court-martial. Speaking for the Royal Navy, Sir James Yeo had been particularly harsh on Prevost during Pring's review. Yeo maintained that had Prevost's troops stormed Macomb's defenses as Downie rounded Cumberland Head, Macdonough's squadron would have been forced to quit the bay. What Prevost should have done, the navy maintained, was to use his vast superiority in numbers—able veterans at that—to take the works across the Saranac and drive the American fleet out from the protection of the bay and into the superior force of the *Confiance* and her consorts.

Prevost certainly had his detractors, but he also had his supporters, among them the Duke of Wellington. "Whether Sir George Prevost was right or wrong in his decision at Lake Champlain is more than I can tell . . . ," the Iron Duke wrote the following December. But, continued Wellington, "I have told the Ministers repeatedly that a naval superiority on the lakes is a *sine qua non* of success in war on the frontier of Canada, even if our object should be solely defensive." Perhaps recognizing that there was more to waging war in the colonies than first met the eye, the Duke of Wellington, it should be noted, turned down an offer to replace Sir George Prevost as governor-general of Canada. Wellington was content to rest upon his Napoleonic laurels until called upon to deliver the coup de grâce at Waterloo the following June.[20]

Prevost's court-martial never convened because the defendant died the evening before it was to commence, no doubt sped to his grave by harsh critics, some of whom—such as Yeo—had little room to talk. Meanwhile, Thomas Macdonough became the toast of the American nation. "In one month from a poor lieutenant I became a rich man," Macdonough himself was to say. "Down to the time of the Civil War," Theodore Roosevelt later wrote, "he [Macdonough] is the greatest figure in our naval history" and the fight that day in Plattsburgh Bay the "greatest naval battle of the war."[21]

Considering the company in which Macdonough found himself, that praise may have surprised even him. But Roosevelt went further. Giving Perry his due for the victory off Put-in Bay, Roosevelt nonetheless wrote: "It will always be a source of surprise that the American public should have so glorified Perry's victory over an inferior force, and have paid comparatively little attention to Macdonough's victory, which really was won against decided odds in ships, men, and metal."[22] Perhaps Thomas Macdonough should have come up with a catchier victory report of his own.

And perhaps history has indeed given Macdonough short shrift. Had not Macomb and Macdonough made their stand, a British general less defensively inclined than Sir George Prevost may well have ended up spending the following winter in Albany if not New York City. At the very least, a British victory at Plattsburgh and on Lake Champlain would have given the Madison administration another flank to worry about while they had their hands full with British dalliances about Washington and Baltimore. And the Union Jack left flying on American soil would have both given credence to Sir George's initial orders to tend to the security of Lower Canada and provided a territorial bargaining chip at the peace table. None of this happened. The gamecock aboard the *Saratoga* had reason to crow. "This is a proud day for America," noted Lieutenant Colonel John Murray of the British army, "the proudest day she ever saw."[23]

ANOTHER CAPITAL BURNS

Lieutenant General Gordon Drummond's efforts at Niagara and General Sir George Prevost's advance to Lake Champlain did not go unsupported by the British government. Indeed, as these incursions were made along the Canadian border, the British lion was angrily raking its claws along the entire Eastern seaboard of the United States. In March 1814 President Madison had finally abandoned his support of the Jeffersonian embargo as a tool of American foreign policy and Congress overwhelmingly repealed the last of the embargo acts. Since 1807 these restrictive trade policies had crippled the American economy far more than the thunder of British cannon, and Americans—particularly Federalists in New England—now hoped that commerce on the seas would flow more freely.

The British response, of course, was predictable. In April 1814 the British blockade of the East Coast was extended from the southern and mid-Atlantic states to cover the coast of New England as well. Boston fumed. Madison denounced the entire blockade as illegal and hoped that neutrals would find renewed trade with the United States more enticing than their fear of the Royal Navy. But with Napoleon in exile, no nation was interested in challenging Britannia's rule of the waves. The consequence was that in 1814 American exports plummeted to one-tenth of what they had been in

1811 and imports shrunk to a quarter of their prewar total. The United States was being commercially strangled.[1]

The chief architect of this strangulation was Vice Admiral Sir Alexander Forester Inglis Cochrane. Born in Scotland in 1758 as the younger son of the eighth earl of Dundonald, Cochrane had literally sailed the world with the Royal Navy, taking part in engagements against the French from the West Indies to Egypt. (Cochrane's nephew, Thomas, was also a daring sailor who became the model for novelist Patrick O'Brian's dashing hero, Jack Aubrey.) In between, Cochrane managed to serve six years in Parliament. After the seizure of Guadeloupe in the Lesser Antilles in 1810, Cochrane was appointed governor of the island, a post he held until succeeding Admiral Sir John B. Warren as commander of the North American station in the spring of 1814. As a full admiral, Warren had commanded all operations throughout North America and the West Indies, but the Admiralty now split these commands, and Cochrane's responsibilities were confined to operations along the Atlantic and Gulf coasts.[2]

As if Cochrane's extension of the blockade to New England wasn't a harsh enough blow to the American economy, the admiral quickly ordered a number of hit-and-run raids along the American coasts an affront to property to be sure, but also a demoralizing assault on the nation's psyche. The larger ports enjoyed some measure of defense from coastal fortifications, but small coastal villages watched in fear for British sails. One British squadron sailed up the Connecticut River and burned twenty-seven ships. The island of Nantucket was cut off from the coast and so ravaged that its inhabitants were forced to declare their neutrality or starve. "Having destroyed a great portion of the coasting craft whose owners were hardy enough to venture to sea," *Niles' Weekly Register* reported, the British "seem determined to enter the little out ports and villages, and burn every thing that floats."[3] And then there was Chesapeake Bay.

• • •

Admiral Cochrane was certainly intent on delivering some measure of retribution all along the American coast, but his determination to do so was intensified by a letter he received from Governor-General Sir George Prevost in July 1814. Remember the unsanctioned aftermath of the American raid on Long Point, Ontario, a few months before? Prevost called upon Cochrane to "assist in inflicting that measure of retaliation which shall deter the enemy from a repetition of similar outrages." Acting upon the request, Cochrane took it upon himself to issue orders to the blockading forces up and down the Atlantic coast to "destroy and lay waste to such towns and districts upon the coast as you may find assailable." Directing his commanders to remember the conduct of "the American army toward his Majesty's unoffending Canadian subjects," Cochrane told them to "spare merely the lives of the unarmed inhabitants. . . ."[4]

The war was taking on a decidedly bitter and vindictive tone. Buffalo for York and Newark, Long Point for Buffalo, Chesapeake Bay for Long Point. However accidental or misguided the American burning of the British government buildings at York had been in the spring of 1813, the end result was a series of escalating reprisals that crossed the line between military operations against military objectives and open warfare on civilian lives and property.

The British had already plundered the Chesapeake Bay country with great impunity during the summer of 1813. Admiral Cochrane now turned to the man who had led those raids, Rear Admiral George Cockburn. If the similarity in names is at first confusing, the irony of the last syllable of Cockburn's name—"burn"—will soon become clear and prevent further confusion. Fourteen years Cochrane's junior, Cockburn had followed a similar Royal Navy career around the globe, from serving as a young lieutenant on Lord Hood's flagship off Toulon in 1793 to sharing in the capture of Martinique from the French in 1809 as part of Cochrane's command. Promoted to rear admiral and given command of a squadron off Cádiz in August 1812, Cockburn was soon on his way to Bermuda to join Admiral Sir John B. Warren's initial blockade efforts along the American coast.[5]

Shortly before Cockburn's arrival there, the *London Times* editorialized on what ought to be done in the conduct of the war. "The paramount duty," the newspaper scolded, ". . . is to render the English arms as formidable in the new world as they have become in the old." To do this, the *Times* went on to conclude, "our plans of hostility . . . ought not to be confined to the North [Canada]. On the contrary, we should contrive as much as is consistent with our own safety, to spare the inhabitants of New York and New England, who are reluctantly dragged into the war. The southern states, on the contrary, as they have most cordially embraced the mischievous politics of the President, ought to receive their richly merited reward."[6] Now, Cockburn was coming to deliver it.

But he was not alone. The British army was also to have a role in this operation. As three brigades of Wellington's veterans were dispatched to assist Sir George Prevost in Canada, a fourth brigade was sent to work in concert with Admiral Cochrane. Its commander was one of Wellington's ablest officers, Major General Robert Ross. Born in Ireland in 1766, Ross was a lifelong veteran of the British Empire's wars. After graduating from Trinity College in Dublin, he was commissioned an ensign in the Twenty-fifth Foot in 1789. As an officer and then commander of the Twentieth Foot, he fought in Holland, Egypt, and Italy, particularly distinguishing himself and his regiment by leading a decisive flanking movement against the French at Maida in Italy's toe. Throughout these campaigns and intervening periods, Ross gained a reputation for indefatigably drilling his troops in every conceivable circumstance. By 1808 the Twentieth Foot made its first deployment to Spain. Four years later, Ross was back in Spain, promoted to major general, and entrusted by Wellington to command a brigade that included his old regiment. That Ross—much like Winfield Scott on the American side—was determined to lead by example is evidenced by the fact that he had two horses shot from under him while his brigade was in the vanguard securing the pass across the Pyrenees at Roncesvalles.[7]

While the British would have cause to second guess Ross's instructions—particularly in light of Sir George Prevost's subsequent, lightning withdrawal from Plattsburgh—Ross was initially ordered "to effect a diversion on the coasts of the United States of America in favor of the army employed in the defense of Upper and Lower Canada." Admiral Cochrane was to determine the point or points of attack, subject to Ross's approval, and Ross was not to conduct "any extended operation at a distance from the coast" or to make any permanent occupation of territory. In other words, hit and run and divert attention from Sir George's Lake Champlain thrust, causing the most havoc in the process. Nothing in Lord Bathhurst's orders to Ross encouraged the destruction of civilian property, or for that matter the burning of nonmilitary public buildings. But his orders did sanction the collection of "contributions in return for your forbearance"—a polite way of saying "demand ransom for their safety."[8] Admiral Cockburn, of course—acting under orders from Admiral Cochrane—was not bound by such civilities and had different ideas.

In fact, George Cockburn seems to have delighted in his reputation for wreaking havoc. Cockburn's previous season on the Chesapeake included a raid all the way north to Havre de Grace at the mouth of the Susquehanna. Afterward, one American, who described himself as a "naturalized Irishman," offered a reward of $1,000 for the admiral's head or "five hundred dollars for each of his ears, on delivery."[9] Thus Cockburn was all too eager to embrace the suggestion of Joseph Nourse, one of his captains. "Jonathan," Nourse reported to Cockburn in a derogatory reference to the Americans, "is so confounded that he does not know when or where to look for us, and I do believe that he is at this moment so undecided and unprepared that it would require little force to burn Washington, and I hope soon to put the first torch to it myself."[10]

On August 15, 1814, the transports carrying Ross's brigade of about thirty-four hundred men passed through the Virginia capes

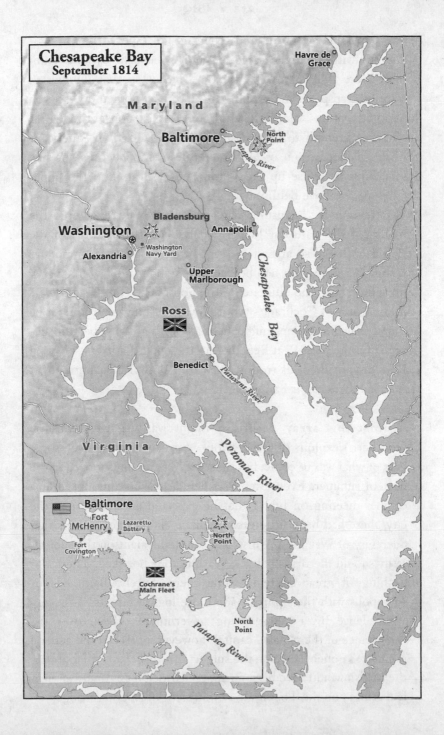

on a favorable east wind and sailed up Chesapeake Bay to rendezvous with admirals Cochrane and Cockburn. By one count there were at least twenty warships and again that number of assorted transports and supply ships in the combined fleet. Seven hundred marines were assigned to join Ross's troops, bringing his effective strength to more than four thousand. The Americans were still not certain of Cochrane's destination because he had sent ships to cruise up the Potomac as well as into the upper Chesapeake above Baltimore.

At noon on August 18 the main British force entered the mouth of the Patuxent River, twenty-some miles north of the Potomac. It did so at first to trap the gunboats of Commodore Joshua Barney, which had been buzzing about the bay like pesky horseflies. But "the ultimate destination of the combined force," in the words of Cochrane's report, "was Washington, should it be found that the attempt might be made with any prospect of success." On August 19, at Benedict on the upper Patuxent, General Ross disembarked his troops and with Admiral Cockburn at his side, led them toward that objective.[11]

Against this array of British might was massed confusion. Cockburn's exploits of the previous year should have given some hint of what was to come, but most of Washington snoozed in the heat of summer. Even after President Madison announced at a cabinet meeting on July 1 that he expected an attack upon the city, there had been little excitement. It didn't help matters that Secretary of War John Armstrong was ambivalent on the subject. Why should the British attack Washington? he asked. There was nothing of great strategic value there. Surely, Baltimore or Annapolis with their ports made more logical targets. This very ambivalence by a high-ranking government official over the importance of the *national* capital, however, mirrored the general populace's nonchalance on the subject of *national* unity. Doubtless the British would have been far more alarmed if an American fleet had cruised up the Thames to within sight of London Bridge.

There were some, of course, who thought that John Armstrong's ambivalence about the importance of Washington stemmed from a more sinister motive. Armstrong's presidential ambitions were well known, and his difficulties with some of his abler generals—chief among them William Henry Harrison—and with fellow cabinet member James Monroe had distinct political overtones. Never a fan of the capital's location, Armstrong, critics contended, might not be too chagrined at its fall. "They [certain critics] think he wishes to have the seat of the government removed, that he may destroy the Virginia combination [Madison-Monroe], which now stands in the way of his promotion to the next presidency." It didn't help matters that it was further reported that Armstrong had rented quarters at Carlisle, Pennsylvania—just in case.[12]

With that background, it wasn't much of a surprise when Madison created the Tenth Military District—essentially Washington, Baltimore, and their environs—that Madison and Armstrong strongly disagreed over the choice of a commander. To his credit, Armstrong favored the military solution, while Madison opted for the political. Armstrong wanted Brevet Brigadier General Moses Porter, currently the commander of the Fifth Military District at Norfolk, Virginia. Almost sixty, Porter had fought in the Revolution, had been with Wayne at Fallen Timbers, and, unlike certain other general officers of that generation, had continued to distinguish himself as the most able artillery officer in the nation's service.

Instead, Madison chose thirty-nine-year-old Brigadier General William H. Winder, a Baltimore attorney of some note whose military exploits left much to be desired. Commissioned a lieutenant colonel in the spring of 1812, Winder had been involved in General Alexander Smyth's stumbles along the Niagara frontier and then stumbled on his own when his troops were surprised and he was captured at the Battle of Stoney Creek the following year. While paroled but not officially exchanged, Winder proved far more capable contributing his legal expertise to negotiations of prisoner exchanges. Madison thought he needed Winder, now officially exchanged, to curry favor with his uncle, Levin Winder, the Federalist governor of Maryland. Surely, Madison thought, Governor Winder and all of Maryland's militia would be quick to rally to his nephew's defense.

The militiamen were critical because, thanks to the brewing engagements at Niagara and Lake Champlain, there were no more than 500 regulars to be had in the entire Tenth Military District. Half of these were cooped up in dubious forts. So Madison issued a massive alert for 93,500 militia to stand ready to be called into active service. On July 12 General William Winder received authorization to call out 6,000 Maryland militia, but Governor Levin Winder's administration responded so slowly that his nephew had fewer than 250 troops under his command six weeks later. So much for blood being thicker than water.[13]

Still, no preparations were made for Washington's defense, although at least a few people were finally growing concerned. "The shameful neglect of the administration to provide an adequate defense for the capital," the *Federal Republican* of Georgetown grumbled, ". . . is a just cause of loud complaint among all parties."[14] Even a neophyte military strategist like William Winder pleaded with Secretary Armstrong that "a thousand determined men might reach the town in thirty-six hours, and destroy it before any general alarm could be given." Armstrong, for his part, miffed because his advice of commander had gone unheeded, chose to ignore Winder and leave him to his own devices. Six weeks later, as Ross and Cockburn disembarked on the Patuxent, Winder was still thrashing about largely on his own.[15]

As the British advanced, General James Wilkinson, awaiting court-martial for the failed St. Lawrence campaign and still the inveterate opportunist, offered his services if his house arrest were to be suspended. But President Madison "concluded that Wilkinson's talents for confusing things were not required in a situation that was already thoroughly confounded and refrained from answering the letter. . . ."[16]

• • •

As Ross and Cockburn moved north from Benedict, a British flotilla shadowed their movements in the Patuxent. In two days the combined forces covered more than twenty miles. The only enemy they encountered was the humid furnace of the beastly hot, dog days of August. "During this short march," Ensign George Robert Gleig of the Eighty-fifth Foot wrote later, "a greater number of soldiers dropped out of the ranks, and fell behind from fatigue, than I recollect to have seen in any march in the Peninsula [Spain] of thrice its duration."[17]

Pausing at Upper Marlborough on August 22, Ross and Cockburn heard the first American fire. It proved to be the distant explosions of Commodore Barney blowing up his gunboats rather than have them be captured. It was a good sign—the Americans were retreating.

From Upper Marlborough, the direct route to Washington led west along the Marlborough Pike and reached the city and the Washington Navy Yard from the south via one of two bridges across the wide mouth of the East Branch of the Potomac River, now called the Anacostia. But surely the capital's defenders would either heavily contest or destroy these bridges. Consequently, Ross and Cockburn opted to feint toward the bridges but then detour on an eight-mile-longer route that crossed the river at a much narrower point. From there, they could descend on the capital from the northeast.

Uncharacteristic of both men, Ross and Cockburn allowed the better part of two days to pass while debating this choice. Marching well into the night of August 23 to make up for lost time, Ross and Cockburn were suddenly overtaken by a messenger from Admiral Cochrane. For reasons not altogether clear, Cochrane had suddenly gotten cold feet and was ordering the entire force back to his ships at Benedict. Ever the good soldier, Ross felt Cochrane's orders left no room for discretion and was inclined to obey. Cockburn argued that they were past the point of return and committed. As field commanders, their decision controlled. Supposedly, the two commanders—who exhibited a high degree of interservice cooperation that was decidedly lacking on both sides in most other engagements of the war—walked in circles through the night and argued about which course to take. By dawn, Cockburn had won out, and the march toward Washington resumed.

In the early hours of August 24, the Americans did the expected and blew up the two bridges across the East Branch of the Potomac. About the same time, the British did the unexpected and turned north for Bladensburg, which was the first easy crossing of the East Branch after it narrowed. Meanwhile, Winder had continued to ride about and organize militia and a handful of regulars. The general concentrated these forces at the navy yard, because he still feared an attack directly from the south, and at Bladensburg. Incredibly, however, as a small British army marched quite leisurely through the Maryland countryside toward the nation's capital, no attempt was made to impede its progress by felling trees, sniping at its flanks, or attacking its supply wagons. Perhaps most significantly, the bridge across the East Branch at Bladensburg was left intact. The upstart American rebels, who had taught the British a thing or two about guerrilla tactics a generation before, seemed to have momentarily forgotten them.

As Ross and Cockburn led the British advance toward Bladensburg without meeting any resistance, Winder was near the navy yard conferring with President Madison and certain members of his cabinet, including Secretary of State Monroe. As it became apparent that Bladensburg would indeed be the battleground, the party started in that direction, with Monroe dashing off well in advance of the others. By the time that Madison, Secretary of War Armstrong, and General Winder arrived above Bladensburg, Monroe had redirected the deployment of some of the militia. Certainly the secretary of state had no direct command authority to do so, and his movement of two units of Baltimore militia so that they no longer supported the front ranks showed that he had no military acumen, either.[18]

Now as noonday temperatures approached 100 degrees Fahrenheit, the advance regiment of Ross's column reached the Bladensburg bridge. Colonel William Thornton's Eighty fifth Foot did not hesitate, but stormed across the bridge in a hail of musket and artillery fire from Winder's front line. President Madison, who had almost blundered onto the bridge moments before, withdrew to watch events unfold from a safer distance. Save for Abraham Lincoln poking his head above the ramparts at the fortifications surrounding Washington during the Civil War, it was the only time that an American president was ever on an actual battlefield as commander in chief. As Thornton's troops came on and Winder's first line retreated, Madison the scholar, the architect of the Constitution, turned to Monroe and Armstrong and advised that "it would be now proper for us to retire to the rear, leaving the military movement to military men."[19]

As Madison and his party started back down the road to Washington, Winder's first line broke and fell back. It probably didn't help matters that as part of his final instructions, Winder had failed to urge a stonewall stand and instead had politely informed his troops that "when you retreat, take notice that you retreat by the Georgetown road."[20] Monroe's misplaced second line held for a time—thanks in part to the well-dressed gentlemen of the Fifth Maryland Volunteer Regiment. But their officers soon decided the odds were against them and ordered a retreat. All along the second line, untrained militia followed suit, and soon what might have been an orderly withdrawal turned into a headlong rout, spurred along by the shrieking of the newfangled Congreve rockets that the British sent streaking through the hot summer sky.

Now Winder was down to a third and final line of defense, placed about a mile above the Bladensburg bridge not by his strategy, but by the happenstance of Commodore Joshua Barney. After destroying his gunboats on the upper Patuxent, Barney was lugging his cannon toward Washington when they were desperately needed. He wasn't in time to entrench his pieces near the river, but instead positioned them across the Bladensburg to

Washington road. As the retreating first ranks fled past his position, Barney and his sailors stood firm and poured a devastating fire into the advancing British column—Thornton's Eighty-fifth Foot still in the lead.

One of Barney's men was Charles Ball, an escaped slave who had served with the commodore's gunboat flotilla for over a year. "I could not but admire the handsome manner in which the British officers led on their fatigued and worn-out soldiers," Ball later wrote. "I thought then, and think yet, that General Ross was one of the finest-looking men that I ever saw on horseback."[21]

Barney's action gave Winder one last chance to rally a defense, and for a time the third line held. But inexplicably, just when Ross and Cockburn's drive might have been blunted—certainly it was being bloodied—Winder ordered the third line to retreat. The chaos of disorganized flight became complete. Elbridge Gerry, Jr., whose father had written President Madison at the war's outbreak that "by war we should be purified," was among those witnessing the rout. "The young ladies," the younger Gerry wrote to his wife, "were very merry relating their attempt to fly on the supposed approach of the enemy to their residence and that they were out run by the militia."[22]

Later some anonymous political wag would characterize this flight back to Washington as the "Bladensburg Races." Washington would not see the likes of it again until the Union rout after First Bull Run in 1861. Recounting Madison's presence there at the start, one stanza of the poem had the president saying:

> *Nor, Winder, do not fire your guns*
> *Nor let your trumpets play,*
> *Till we are out of sight—forsooth,*
> *My horse will run away.*[23]

Truth be known, the sixty-three-year-old Madison had been ill that summer and displayed ample personal courage and stamina by riding a circuit of almost twenty miles before returning to his residence in the city to check on the safety of his wife. For Commodore Barney's part, he and many of his sailors stood by their guns and were captured by the advancing British. Despite the American retreat, Ross's men had paid dearly for the ground. At least 64—and by one account as many as 180—British troops were killed, hundreds wounded. American losses were estimated—again, with varying figures—as only 26 killed and 51 wounded. The victors paused in the late afternoon heat to catch their breath. They made no immediate attempt to pursue the Americans, because, as Cockburn reported to London, "the victors were too weary and the vanquished too swift."[24]

As evening fell, a brigade of Ross's command, which had not been engaged in the Bladensburg fight, passed through the other troops and, with General Ross and Admiral Cockburn in the advance guard, marched down the road that led to Washington, seven miles distant. It was twilight by the time Ross and Cockburn rode rather boldly down Maryland Avenue and into the square on the east side of the Capitol. Suddenly a volley of shots rang out from the large house on the northwest corner of Second Street, Northeast, and Maryland Avenue, for some time the residence of Secretary of the Treasury Albert Gallatin. Early in the summer Gallatin had written to Monroe from London that the British, "to use their own language, . . . mean to inflict on America a chastisement that will teach her that war is not to be declared against Great Britain with impunity."[25]

Now Gallatin got a personal taste of that sentiment. The volley killed one British soldier and wounded three others. General Ross, too, fell to the ground. Ross, who had already lost one mount earlier in the day at Bladensburg, had just had another horse shot from beneath him. Rumors would abound for years about who fired the shots, but Ross ordered Gallatin's house burned, one of the few private dwellings so destroyed. Variations would also persist about whether Ross and Cockburn were advancing under a flag of truce, attempting to negotiate for the city's surrender to avoid more bloodshed. As it was, there were no representatives of the American government left with whom to negotiate. Within minutes, the Capitol, too, was ablaze.[26]

After departing Bladensburg, President Madison had returned to the President's House—as the White House was then called—to inquire after his wife. Brave Dolley Madison had already loaded what state papers she could into a carriage and ensured herself a place in legend by saving Gilbert Stuart's painting of George Washington. "This process [of removing it] was found too tedious for these perilous moments," she took time to write her sister, "[so] I have ordered the frame to be broken, and the canvas taken out."[27]

Born in Virginia, Dolley Payne Todd Madison was forty-six. Widowed at twenty-five from Quaker lawyer John Todd, she married the bookish Madison, seventeen years her senior, in 1794. When he became Jefferson's secretary of state in 1801, Dolley Madison frequently served as the widowed president's hostess at official functions. Dolley and James Madison moved into the President's House in their own right in 1809. She continued the gracious hostess role for which she is most remembered, but may well have also held her quiet husband's ear on matters of policy and patronage. If her later words are to be believed, Dolley Madison would have fought the British all the way down Pennsylvania Avenue. "If I could have had a cannon through every window," wrote Dolley, "but alas! those who should have placed them there, fled before me, and my whole heart mourned for my country!"[28]

So, with the Capitol in flames, Ross, Cockburn, and an advance guard rode northwest down deserted Pennsylvania Avenue to the President's House. It, too, was deserted. "So unexpected was our entry and capture of Washington," Ross reported, that "when our advanced party entered the President's house, they found a table laid with forty covers." It had been a long day, and the British helped themselves to the feast of abundant food and wine, toasting the Prince Regent and damning Madison in the process. Then, after pillaging the Madisons' personal possessions on the second floor,

British sailors set multiple fires throughout the building. "Our sailors were artists at the work," one of the Royal Navy captains recalled. That sailors did the deed suggests that General Ross may have been feeling a bit uncomfortable by now. Cockburn, for his part, appears to have joined in both the toasts and looting.[29]

As flames roared out of the windows of the President's House, the invaders moved on to the nearby Treasury Building. It, too, was soon in flames. By now, these conflagrations lit up the August sky, but an even larger fire was burning to the south. On orders from Secretary of the Navy Jones, the Americans had set fire to the Washington Navy Yard to prevent its warehouses and munitions from falling into British hands. Also put to the torch were the hulls of the frigate *Columbia* and the sloop of war *Argus* that were under construction.

After the Treasury Building was ablaze, the British retired back up Pennsylvania Avenue to an encampment on Capitol Hill. Ross and Cockburn stayed at the home of physician James Ewell at First and A Streets, Southeast, where the Library of Congress now stands. The British rank and file were exhausted, but few citizens who remained in Washington slept that night. "You never saw a drawing room so brilliantly lighted as the whole city was that night," wrote Mary Hunter as she watched the navy yard and buildings burn. "Few thought of going to bed—they spent the night in gazing on the fires and lamenting the disgrace of the city."[30]

The following morning, British troops marched to the State and War Department buildings west of the now gutted President's House and also put them to the torch. Only the timely appeal of Dr. William Thornton, the superintendent of patents, saved the Patent Office from a similar fate. Meanwhile, Admiral Cockburn was settling an old score. At the offices of the *National Intelligencer*, halfway between the Capitol and President's House, Cockburn watched as the presses and types of the newspaper were wrecked. The admiral lamented that publisher Joseph Gales was nowhere to be found, but supposedly declared, "Be sure that all the C's are destroyed so that the rascals cannot any longer abuse my name."[31]

As Cockburn oversaw the destruction of the *National Intelligencer*, there was still no armed resistance, but if one report is to be believed, there was at least one taunt. An American onlooker is supposed to have shouted at Cockburn that "if General Washington had been alive, you would not have gotten into this city so easily." "No sir," the admiral is supposed to have responded. "If General Washington had been president we should never have thought of coming here."[32]

By afternoon the chaos of the British occupation was almost complete, when a sudden and violent thunderstorm struck. The skies darkened almost more so than the night before, and lightning darted from black clouds to the ground. For those who believed in omens, it was a scene rife with poignancy. One account, undoubtedly sensationalized long after the fact, has a woman stepping out of her house to confront Admiral Cockburn as he exclaimed, "Great God, Madam! Is this the kind of storm to which you are accustomed in this infernal country?" To which the woman replied, "No, sir, this is a special interposition of Providence to drive our enemies from our city."[33]

Doubtless such predictions had little influence on Cockburn, but both he and Ross knew that it was time to go. Ross seems to have been particularly anxious to retreat before the American militia could regroup. On the evening of August 25, barely twenty-four hours after their arrival, Ross gave the order to march back to Bladensburg and from there posthaste to Benedict and the waiting British ships. To enforce an 8:00 P.M. curfew on the town's inhabitants, Ross ordered huge bonfires lit to create the illusion that the British were indeed still in their encampment. By dawn, the exhausted British troops were sleeping in heaps well beyond Bladensburg.[34]

Scarcely had Ross's troops marched east from Capitol Hill than the British navy appeared in the Potomac River on the other side of town. Admiral Cochrane had dispatched Captain James A. Gordon in command of the frigate HMS *Sea Horse* and a small squadron to sail up the river to protect Ross's flank and extricate his troops should they be cut off from a return to the Patuxent. This did not prove the case, but as long as Gordon was in position, he took Fort Washington without a fight and tied up at the wharves of Alexandria. There his sailors seized a number of small vessels and loaded them with 16,000 barrels of flour, 1,000 hogsheads of tobacco, 150 bales of cotton, and assorted tar, beef, and sugar. That the town itself was saved from destruction was due to its surrender and acquiescence in this shopping spree.

By now a determined Dolley Madison had returned to the President's House despite the presence of the British ships. "I remained nearly three days out of town," she wrote, "but I cannot tell you what I felt upon re-entering it—such destruction— such confusion! The [British] fleet in full view and in the act of robbing Alexandria!"[35] Gordon's squadron finally cast off on September 1, and sailed back down the river to rejoin Cochrane.

So the invading British were gone from Washington—gone almost as quickly as Sir George Prevost and his legions had disappeared before Plattsburgh. But the ramifications of the smoldering ruins of the capital city that the resolute Ross and cavalier Cockburn left behind them would echo far more profoundly. "Certain it is," remarked a stunned *New York Evening Post*, "that when General Ross's official account of the battle and the capture and destruction of our capitol is published in England, it will hardly be credited by Englishmen. Even here it is still considered as a dream."[36]

Officially, an exasperated James Monroe complained to Admiral Cochrane that "in the course of ten years past, the capitals of the principal powers of the continent of Europe have been conquered, and occupied alternately by the victorious armies of each other; and no instance of such wanton and unjustifiable destruction has been seen."[37]

American anger was not restricted to the British. There was plenty of blame to be cast about. "George Washington founded this city after a seven years' war with England," read graffiti on the scorched walls of the Capitol; "James Madison lost it after a two years' war."[38] For Secretary of War Armstrong, it was the final blow. "Armstrong's management of the northern campaign caused severe criticism," Henry Adams was to write eighty-some years later, "but his neglect of the city of Washington exhausted the public patience."[39]

Madison graciously advised Armstrong to get out of town for a while until tempers cooled, but Armstrong saw the handwriting on the wall. On September 27 he resigned, blaming his own fall on Monroe. "I was supposed to be in somebody's way [for the presidency]," wrote Armstrong, "and it became a system to load me with all the faults and misfortunes which occurred."[40] It was hardly that, but when Madison named Monroe acting secretary of war as well as secretary of state, Armstrong could claim his paranoia justified.

Next on the list was General Winder. For more than six weeks before the British landed, Winder had scurried about without much effect. One can only ponder what the outcome at Bladensburg might have been had an equal of Ross—someone such as Winfield Scott, for example—been given time to organize the militia and impede the British advance. Scott was then recuperating from his wounds at Lundy's Lane, but by coincidence, he would preside at Winder's court of inquiry. No doubt taking into account the lack of support from the administration in general and Armstrong in particular, as well as the vexing issue of command and control of independent militia, the court found Winder "entitled to no little commendation, notwithstanding the result. . . ."[41]

The New York Evening Post's claim notwithstanding, reaction on the other side of the Atlantic was mixed. "The Cossacks spared Paris," the London Statesman declared, "but we spared not the capitol of America." Some members of Parliament were equally aghast, but the Prince Regent termed the campaign "brilliant and successful."[42] And even as he departed Canada to face his own court of inquiry over his conduct at Plattsburgh, Sir George

Prevost professed that "as a just retribution, the proud capital at Washington has experienced a similar fate to that inflicted by an American force on the seat of government in Upper Canada."[43]

Some years later, in recapping effective maritime expeditions of the early nineteenth century, the great military strategist Antoine Henri Jomini called the British raid on Washington extraordinary. "The world was astonished to see," wrote Jomini, "a handful of seven or eight thousand Englishmen making their appearance in the midst of a state embracing ten millions of people, taking possession of its capital, and destroying all the public buildings—results unparalleled in history."[44] But this experience, too, would go into the pot that was slowly boiling the glue that would come to bind together a nation.

O Say, Can You See?

Buoyed by stunning success at Washington, the British pondered their next move. Less than forty miles to the northeast, Baltimore, queen city of the Chesapeake, emitted a siren's song of invitation. Captain Gordon's plunder snatched from the docks of Alexandria paled in comparison to the riches lining Baltimore's waterfront. One had only to look at Washington's shabby defense to realize that to many it was a national capital in name only. Baltimore, on the other hand—its forty-five thousand residents making it the third largest city in the United States—was the acknowledged commercial hub of the mid-Atlantic states. An attack against Baltimore would be an attack against the American jugular.

In addition to its commercial value, there were also military and political motivations for advancing against Baltimore. Its fine harbor had long provided shelter for oceangoing privateers as well as Commodore Joshua Barney's flotilla of gunboats. Politically, the city was about as pro-Republican, prowar, and anti-British as was possible. A lightning thrust here would indeed strike deep into the hornet's nest. "I do not like to contemplate scenes of blood and destruction," Admiral Cochrane's fleet captain, Edward Codrington, wrote to his wife, "but my heart is deeply interested in the coercion of these Baltimore heroes, who are perhaps the most inveterate against us of all the Yankees."[1]

All these factors made Baltimore a tempting target, but Cochrane's high command was divided on a course of action. At first blush, Cochrane himself seems to have been inclined to escape the caldron of late-summer heat and concomitant malaria and yellow fever on the Chesapeake and sail to more friendly Rhode Island to regroup. With refreshed troops and suitable reinforcements, he could yet descend upon Baltimore—or perhaps Charleston or Savannah—as he made his way south toward the Gulf Coast later in the fall.

Not surprisingly, Admiral Cockburn, ever the brash swaggerer, was of a different mind. His fleet and General Ross's troops had just proven themselves the better of the American militia. Their British forces were not only still fit, but also in the immediate vicinity. Why come back later to do a job that might be done more easily *now*—before the demoralized Americans were given time to assemble regulars. That left General Ross somewhere in the middle between the two naval officers.

One strongly suspects that Ross retreated from the fires of Washington decidedly less enthused over the operation than either Cockburn or Cochrane. Perhaps he even carried with him some measure of embarrassment. This was not the type of warfare Ross had practiced under Wellington in Spain. Indeed, Cochrane reported Ross's apparent lack of enthusiasm for the Royal Navy's campaign of vengeance shortly after Ross's troops reembarked at Benedict. Cochrane didn't doubt that Baltimore "ought to be laid in ashes" at some point. But "if the same opinion holds with His Majesty's ministers," Cochrane wrote to the First Lord of the Admiralty, "some hint ought to be given to General Ross, as he does not seem inclined to visit the sins committed upon His Majesty's Canadian subjects upon the inhabitants of this state."[2] In other words, the Royal Navy thought that the general who always led in the vanguard of his troops was a little "soft."

For his part, General Ross was dubious about the military objectives to be gained by attacking Baltimore. Commercial plunder aside, Ross thought that at best a successful campaign against

Baltimore would be anticlimatic after the burning of Washington, and at worst, well, the Americans were proving elsewhere that they could fight. What if his force was somehow cut off from Cochrane's fleet? If that happened, Prevost would not be the only one worrying about another Saratoga.

In short, Ross saw little upside in the venture and a great deal of risk. "May I assure Lord Bathurst you will not attempt Baltimore?" his aide, Captain Harry Smith, asked the general as Smith left with dispatches for England aboard HMS *Iphigenia*. "You may," Ross replied decidedly.[3] But Cockburn was an arm twister. When unfavorable winds delayed the fleet's departure for Rhode Island, the continuing debate among the senior staff was finally resolved. On September 7, 1814, Admiral Cochrane made the decision to attack Baltimore—immediately.

Four days later, after Captain Gordon's Potomac flotilla had rejoined the main fleet, Cochrane's ships anchored in the wide mouth of the Patapsco River about ten miles from the heart of Baltimore. If Washington had been ill-prepared for an attack, Baltimore was just the opposite. On the day of the "Bladensburg Races," Baltimore's town fathers had formed a "Committee of Vigilance and Safety," which promptly drafted every white male between sixteen and fifty for duty. Elderly men "who are able to carry a firelock and willing to render a last service to their country and posterity" were invited to form their own company. By the time that three American cannon boomed notice of the British fleet's arrival in the Patapsco, about nine thousand militiamen were assembled in and around the city, busily adding to an elaborate system of earthworks that had been in preparation for over a year. Whereas Washington ran, Baltimore dug.

The architect of Baltimore's civilian defense was Samuel Smith, a major general of the Maryland militia and one of the state's United States senators. Born in 1752 to one of Baltimore's well-established merchant families, Smith counted military experience going back to the early days of the Revolution and political experience dating from his first election to the House of Representatives in 1792. At sixty-two, Smith had a depth of experience and was not about to be pushed around by anyone—British or American. Thus, when William Winder came galloping into town, fresh from his rout at Bladensburg and waving his regular army commission as a brigadier general, *volunteer* Major General Smith said thanks, but no thanks. Smith didn't care who young Winder's uncle was or what his regular army rank. He stoutly, and shrewdly, refused to relinquish his command. In fact, Governor Levin Winder seems to have agreed with Smith's decision and proceeded to move the remainder of the state's militia to Baltimore with far greater haste than he had done to Washington.[4]

Central to Baltimore's watery defenses was Fort McHenry. Named for John Adams's secretary of war, the fort had been built in the 1790s as part of a system of coastal defenses. Typical of fortifications of that era, it was constructed of masonry and dirt fill in the shape of a five-pointed star, with five bastions overlooking an outer perimeter. Fort McHenry sat at the tip of a stubby peninsula dividing the Northwest Branch of the Patapsco River and Baltimore's secluded inner harbor from the western or Ferry Branch of the river. With a three-gun battery opposite the fort at Lazaretto on the eastern bank of the Northwest Branch and smaller Fort Covington and another six-gun battery located about a mile west of McHenry, any invading force sailing directly into the upper Patapsco was apt to receive a warm welcome.

Admiral Cochrane knew this, but also knew that an even greater defense was the shallow depth of the river that precluded his larger vessels from sailing right up to the fortifications. Accordingly, Cochrane disembarked Ross and some four thousand soldiers at North Point at the mouth of the Patapsco. Ross was ordered to advance on Baltimore via land from the east while Cochrane's shallower draft vessels shadowed their movements from the Patapsco, much as they had done a few weeks before on the Patuxent en route to Washington. Once again, Admiral Cockburn insisted on accompanying Ross and his troops.

• • •

Left in the care of the main British fleet was a small American ship, the *Minden*, which had sailed from Baltimore the week before under a flag of truce to seek an audience with Admiral Cochrane. Aboard the *Minden* were a Georgetown lawyer named Francis Scott Key and Baltimore lawyer John S. Skinner, the latter an agent of the American government appointed to negotiate prisoner exchanges. Skinner's present objective was the release of a sixty-five-year-old country doctor named William Beanes.

At first glance, Beanes appeared to be an unlikely prisoner. Born in Prince Georges County, Maryland, of Scottish immigrants, Beanes practiced medicine and farmed, eventually becoming one of the county's major landowners and most respected citizens. When General Ross and Admiral Cockburn marched through Upper Marlborough en route to Washington, they found the town largely deserted except for the home of Dr. Beanes. The good doctor offered the British officers the use of his house on the night of August 22. Both Ross and Cockburn departed the following morning thinking that the gracious gentleman who still spoke with a Scottish brogue and freely espoused his Federalist leanings was something of a friend to His Majesty's cause.

Beanes was hardly that. After the British column passed through Upper Marlborough again four days later on its retreat back to the fleet, Beanes helped detain six stragglers as prisoners of war. But the British were not quite gone. Early the next morning a mounted detachment of British troops appeared at the Beanes home, unceremoniously hauled Beanes and two houseguests out of bed, and hustled all three off to the British fleet at Benedict, a ride that left the aged Beanes jostled at the very least. According to the *Baltimore Federal Gazette*, their captors left word in Upper Marlborough that unless the six British prisoners were released, the British would return and burn the town. The release was promptly arranged. Subsequently, Beanes's houseguests, fellow physician Dr. William Hill and Philip Weems, neither of whom appear to have taken any role in the capture of the British soldiers, were released as well. Dr. Beanes, however, was a different matter.

Both Ross and Cockburn appear to have taken Beanes's conduct in the soldiers' capture as personal affronts. They now saw his prior hospitality as deceitful at best, treasonable at worst should his Scottish accent disprove his claim of Maryland birth and make him a British subject. Consequently, Beanes was detained aboard the British fleet with no apparent chance of release. Fearful that the doctor might be bound for Halifax or Bermuda to stand charges of goodness knows what, his neighbors in Prince Georges County quickly rallied to his defense. They did what some people still do; they called in a Washington lawyer.

Francis Scott Key was well connected. He had been practicing law in Georgetown since 1802. His older sister was married to one of Dr. Beanes's most affluent patients, and it was her husband, Richard E. West of Woodyard, who rode to Georgetown to ask Key's assistance. Key promised to do what he could and set off for Baltimore to meet up with John Skinner. En route, Key stopped in Washington and Bladensburg and shrewdly obtained letters from wounded British soldiers recuperating in makeshift hospitals there that attested to the good treatment they were receiving from their American captors.

When the *Minden* finally pulled alongside Admiral Cochrane's flagship, HMS *Tonnant*, a week later, the two lawyers were treated cordially by the admiral and invited to dine with him and Cockburn and Ross. But the cordiality turned cool when the purpose of their visit focused on Dr. Beanes. Cockburn went into one of his rages and was inclined to deny the lawyers access to Beanes even for the purpose of delivering soap and fresh underwear. Ross was of an equal mind until Key played his trump card. Ross read several of the letters from his wounded soldiers, turned moody, and then gradually softened. Only based on his appreciation for the care of his men, the general said, would he acquiesce in the doctor's release. Dr. Beanes was the army's prisoner, and over Cockburn's objections, the army would release him, but *not* until it had finished its business at Baltimore.[5]

So, as Ross disembarked with his men at North Point at 3:00 A.M. on September 12, Key and Skinner were reunited with Dr. Beanes aboard the *Minden* and left under the guns of Cochrane's main fleet. At first, Ross's fourteen-mile march to the city resembled the British advance on Washington. General Samuel Smith, for all his preparations, had been certain that the British attack would come by water and had dug his breastworks accordingly. But there were early signs that this advance would be different. "The Americans had at last adopted an expedient which, if carried to its proper length," Ensign George Robert Gleig of the Eighty-fifth Foot reported, "might have entirely stopped our progress. In most of the woods they had felled trees, and thrown them across the road; but as these abattis were without defenders, we experienced no other inconvenience than what arose from loss of time. . . ."[6]

By 8:00 A.M., however, the first brigade of Ross's advancing column met with more than inconvenience. Between Bear Creek and the Back River, where the North Point peninsula was only about a mile wide, Smith spread out more than three thousand Baltimore militia under the command of Brigadier General John Stricker. Just in advance of this American line, Ross and Cockburn stopped for breakfast at the Gorsuch farm. When asked by their reluctant host if they would be returning for supper, legend has Ross replying, "No, I'll eat in Baltimore tonight—or in hell."[7] A short time later, a hail of musket fire poured into the British column from the nearby wood.

Ross quickly surmised that they had stumbled onto more than just a handful of militia. Leaving Cockburn with the first brigade, Ross himself galloped back down the road to hurry the advance of the rest of his column. As he did so, a lone shot rang out from the trees surrounding the road. The general who always led by example was struck by a bullet that passed through his right arm and into his chest. Reports vary as to who was with Ross when he fell and how long he may have lain in the road. Various accounts have him falling into the arms of an aide, lying alone until found by approaching troops, ordering up his second in command, and/or breathing his last words about his wife.

Only six months before, Elizabeth Ross had ridden over the snowy mountains of the Pyrenees from Bilbao, Spain, to St. Jean de Luz, France, to nurse her husband after he was severely wounded at Orthes. Upon departing for America, Ross promised her that it would be his last campaign. It was. Elizabeth Ross outlived her husband by more than thirty years, taking little comfort that a royal warrant had decreed that she and her descendants be henceforth known as Ross of Bladensburg.[8]

Admiral Cockburn hurried to Ross's side as he lay dying. Differences in personal temperament and interservice rivalries aside, Ross and Cockburn had worked extremely well together in a war that had seen army and navy on both sides thwart each other as much as the enemy. Now that cooperation waned. "Our country has lost in him one of its best and bravest soldiers," avowed Cockburn, "and those who knew him, as I did, a friend most honoured and beloved."[9] After the battle, Ensign Gleig would pen a junior officer's assessment: "The death of General Ross seemed to have disorganized the whole plan of proceedings, and the fleet and army rested idle, like a watch without its main spring."[10]

Command of the column now devolved upon Colonel Arthur Brooke, and by midmorning Brooke ordered his eight artillery pieces to pound the American line. Admiral Cockburn rode up and down the British line on a white charger, conspicuous in his gold-braided hat. It was almost as if he was daring the Americans to hit him. Grieving for their fallen general, the British troops were nonetheless able to joke about Cockburn's presence, warning each other to stay "out of the admiral's way because he drew so much fire. 'Look out my lads! Here is the admiral coming! You'll have it directly!'"[11]

As the British infantry advanced, Stricker's militia fell back to second and third lines of defense, but unlike Bladensburg, their withdrawal was orderly and the third line held. By nightfall the British had won the field of battle, but at a stiff cost. And they were not much closer to entering Baltimore itself than they had been after Ross's breakfast boast. In addition to hundreds of casualties on both sides, two American sharpshooters, one of whom likely fired the fatal bullet that struck Ross, were dead. Other American sharpshooters were given no quarter. The British did not fight from trees.[12]

Admiral Cochrane, too, grieved Ross's death, but knew that something had to be done quickly to support Brooke. At dawn on the following morning, September 13, the admiral dispatched five bomb ketches capable of hurling two-hundred-pound mortar shells four thousand yards and the rocket vessel HMS *Erebus* to take up station less than three miles below Fort McHenry. His hope was to reduce the fort and outflank the American position on the peninsula, opening the way into the city for Brooke. The larger British ships, hampered by their deeper drafts, waited down the river.

Fort McHenry's commander was thirty-four-year-old Major George Armistead. A Virginia native, Armistead came from a long family tradition of military service and had made the army his career, transferring to the artillery after a stint with the infantry. He was with Winfield Scott at the capture of Fort George before his appointment to command Fort McHenry early in 1814. Aside from his professional concerns at the moment, Armistead was filled with plenty of personal anxiety. His wife, Louisa, was due to deliver their first child any day and had taken refuge with family in Gettysburg, Pennsylvania. "I wished to God you had not been compelled to leave Baltimore," Armistead wrote her two days before Ross landed at North Point. "I dreamed last night that you had presented me with a fine son. God grant it may be so and all well."[13]

High above Fort McHenry an American flag stood out in the morning light. By any standard, it was a huge ensign. Thirty feet tall and forty-two feet wide, the flag boasted fifteen five-pointed stars arranged in five staggered rows of three stars each on a blue field set on fifteen stripes of alternating red and white. The fifteen stars and stripes were for the original thirteen states and the additions of Vermont and Kentucky. The newer states to join the Union—Tennessee, Ohio, and as recently as 1812, Louisiana— were not yet accorded the honor of a star. (In 1818, Congress authorized a flag with thirteen stripes and one star for each state to be added on the Fourth of July following its admission.) Legend has it that Armistead, or perhaps the fort's prior commander, asked Mrs. Mary Young Pickersgill, a flagmaker for Baltimore's merchant fleet, to sew a flag so big that the British would have no trouble finding the fort. She did so with the aid of her thirteen-year-old daughter, Caroline, working by candlelight in a nearby brewery in order to spread out the voluminous yards of cloth.[14]

Now the British had indeed found Armistead, and despite his twenty cannon at Fort McHenry and the flanking batteries at Lazaretto and Covington, there was little that Armistead could do to answer the hail of British mortars and rockets that pounded his positions throughout the day on September 13. It was a matter of simple math. The range of the British mortars was four thousand yards; the range of his cannon was two thousand yards. All through the day and into the night of September 13, the British ships kept up their bombardment.

Downstream aboard the anchored *Minden*, Francis Scott Key, John Skinner, and the soon-to-be free Dr. William Beanes could only pace the deck and wonder what havoc this bombardment was wreaking. Only the sight of Mary Pickersgill's huge flag waving above the harbor gave some assurance that the fort had not fallen, that Baltimore was secure. Then, darkness fell. But the bombardment continued. The contrails of the Congreve rockets from the *Erebus* glared red across the sky. The mortars, with their fuses timed to explode above the fort and rain shrapnel and destruction upon it, appeared as giant fireworks—bombs bursting in the night sky.

Dr. Beanes was beside himself. He kept asking his companions over and over again, "Can you see the flag? Is the flag still there?"

Occasionally Key could get enough of a glimpse through a telescope to assure the doctor that the flag was indeed still there. But then the American guns in the fort opened fire. Admiral Cochrane was trying an end run, rowing troops in the darkness past the fort to attack Fort Covington and the western batteries in an attempt to outflank it. Meanwhile, Brooke was supposed to advance against Stricker's last line of defense to the east. Armistead's batteries fired away at the British rowboats and then fell silent. Had the British forces been repelled, or had the fort fallen?

Once again, Dr. Beanes asked his anguished question: "Was the flag still there?" Finally, as the early light of dawn tinged the eastern sky, Key was able to answer in the affirmative. What so proudly they had hailed at the twilight's last gleaming was indeed still there. Beanes and Skinner were jubilant. Key took out an envelope and scribbled a few lines. Two days later, back in an American Baltimore, Key elaborated on his verse.

> *O say, can you see, by the dawn's early light,*
> *What so proudly we hail'd at the twilight's last gleaming?*
> *Whose broad stripes and bright stars, thro' the perilous fight,*
> *O'er the ramparts we watch'd, were so gallantly streaming?*
> *And the rockets' red glare, the bombs bursting in air,*
> *Gave proof thro' the night that our flag was still there.*[15]

Fort McHenry had held. Major Armistead's troops had suffered only four killed and twenty-four wounded. Soon news would come from Gettysburg that the major was the father of a healthy newborn—a girl. George Armistead would die young four years later, but his family would have cause to focus on Gettysburg again a generation hence. His nephew, Lewis Armistead, would die there at the head of his troops during Pickett's Charge on a hot day in July 1863.

Meanwhile, as Cochrane's forces failed to reduce Fort McHenry, Colonel Brooke's attempt to force Stricker's line had also failed—stillborn in the rainy early hours of September 14. Cochrane had warned Brooke not to storm the American earthworks unless he was certain of carrying them. With Cochrane's promise of the full support of the navy from the Patapsco side appearing dimmer and dimmer as night edged toward dawn, Brooke reconsidered the odds. "Under these circumstances," Brooke reported to Cochrane, "and keeping in view your Lordship's instructions, it was agreed between the Vice-Admiral [Cockburn] and myself that the capture of the town would not have been a sufficient equivalent to the loss which might probably be sustained in storming the heights."[16]

Two days before, some four hundred miles to the north, General Sir George Prevost, with three times the force arrayed against his opponents, had come to the same conclusion. Like Ross departing Washington, Brooke left his campfires burning brightly and slipped quietly away to return to Cochrane's fleet. The British campaign against the Chesapeake was over. The next day, under favorable winds, the British fleet began its departure from the bay. Admiral Cochrane sailed for Halifax; cocky Cockburn headed for Bermuda; the bulk of the troop transports marked time until mid-October when they sailed for Jamaica to await a grand rendezvous and Cochrane's next thrust along the Gulf Coast.

In their wake, the British left mostly bitterness. Rabid Republican Hezekiah Niles of the *Niles' Weekly Register* went so far as to propose that a monument be erected at the spot where Ross fell, emblazoned with the following inscription. "By the Just Disposition of the Almighty near this Spot was Slain, September 12, 1814, the Leader of a Host of Barbarians who destroyed the Capitol of the United States . . . and devoted the Populous City of Baltimore to rape, robbery, and conflagration."[17] It didn't happen quite that way, of course, and if anyone deserved the sentiment, it was Admiral Cockburn and not the deceased Ross.

Other words were more enduring. Joseph H. Nicholson was instrumental in publishing Francis Scott Key's poem in the form of a handbill entitled "The Defense of Fort McHenry." It was so well received that it was republished by a newspaper in Frederick, Maryland, under the title "The Star-Spangled Banner." Then, barely a month after the event, the words were sung publicly for the first time to the tune of an old English drinking song, "To Anacreon in Heaven."[18]

Over the years "The Star-Spangled Banner" gained in popularity, until in 1931 Congress finally declared it America's national anthem. Every schoolchild learned its first verse. Those who looked beyond the familiar words to the anthem's second verse may have never known that the "dread silence" of Armistead's guns for so long had been due to their short range.

> On the shore dimly seen thro' the mists of the deep,
> Where the foe's haughty host in dread silence reposes,
> What is that which the breeze, o'er the towering steep,
> As it fitfully blows, half conceals, half discloses?[19]

And those who made it to the fourth verse and knew the war's history may have thought Francis Scott Key patriotically prophetic, but pragmatically premature when he declared:

> Blest with vict'ry and peace, may the Heav'n-rescued land,
> Praise the Pow'r that made and preserved us a nation![20]

The war was still far from won.

STILL MR. MADISON'S WAR

As the smoke cleared from Washington and the British fleet departed Baltimore, the question on most Americans' minds remained, now what? Despite Prevost's withdrawal from Lake Champlain and an uneasy stalemate along the Niagara, Admiral Cochrane's fleet was sailing south to nip at yet another flank of the young republic. There was tremendous anxiety and uncertainty. More than two years of conflict had failed to wash away the reluctance to fight that many had felt at the outset. New England, in particular, was still an ambivalent partner in the war effort. The excitement with which Boston and its environs had once celebrated the victories of "Old Ironsides" over the *Guerrière* and *Java* had long since faded in the gloom of the ever-tightening British blockade.

Despite this widespread sentiment, in October 1814 Secretary of War James Monroe renewed the call for pushing the war into Canada, claiming that "this was the best way to secure the friendship of the Indians, protect the coast, and win peace." Later, Monroe wrote to Major General Jacob Brown that "the great object to be attained is to carry the war into Canada, and to break the British power there, to the utmost practicable extent."[1]

Clearly, John Randolph's whip-poor-will was still cawing incessantly, but where, oh where, had Monroe been during the fizzled campaigns of the last three years? Even Major General

George Izard, who had been hastily dispatched from Plattsburgh to the Niagara frontier just before Prevost's approach, was forced to concede to Monroe a high degree of impotence on the issue of Canada. "I confess, sir, I am embarrassed," Izard wrote the secretary after arriving at Niagara. "At the head of the most efficient army which the United States has possessed during this war, much must have been expected from me—and yet I can discern no object which can be achieved at this point, worthy of the risk which will attend its attempt."[2]

But even breathing the word "Canada" made most Federalists see red. They had been saying from the outset that the war had been declared and was being waged for territorial gains and not in the interest of resolving disputes upon the high seas. Truth be told, of course, the Madison administration had indeed far more pressing matters with which to contend in the fall of 1814 than Canada. When President Madison summoned Congress into session in Washington on September 19—about a month earlier than planned—there was even considerable doubt that the burned-out city would remain the nation's capital. The destruction that welcomed members of Congress only underscored the dismal sentiments many were hearing in their home districts. This last session of the Thirteenth Congress met in the Patent Office, which some congressmen were quick to call "confined, inconvenient, and unwholesome."[3]

The chief concern was the economy. "We are in a deplorable situation," Assemblyman Daniel Sargent told the Massachusetts General Assembly, "our commerce dead; our revenue gone; our ships rotting at the wharves. . . . Our treasury drained—we are bankrupts."[4] Not only had New England's coasts felt the tread of British raiding parties, but the British had occupied Maine east of the Penobscot River and extorted Nantucket Island to the point that it almost became a British port of call. There was almost no national presence along the coast for defense, and what defensive measures were undertaken were done so by the individual states only at their own expense and over their loud objections.

Things weren't much better in the south. From the Chesapeake country southward, commodity prices plummeted as exports of cotton, tobacco, wheat, and other agricultural products sat bottled up on wharves, compliments of the British blockade. Even Thomas Jefferson wrote to fellow Virginian William Short complaining about the abundance of crops. "For what can we raise for the market," Jefferson asked Short rhetorically. "Wheat? We can only give it to our horses, as we have been doing ever since harvest. Tobacco? It is not worth the pipe it is smoked in. Some say whiskey; but all mankind must become drunkards to consume it."[5]

And the only reason that Pennsylvania, New York, and the West were not in similar straits, but were in fact enjoying some degree of prosperity, was that the federal government was pouring millions of dollars into their economies to build roads and military installations and to feed and clothe troops. It was not lost on folks in other parts of the country that many in the west who had championed the war were now reaping its rewards. "The warhawks," grumbled the Federalist *United States Gazette*, "are thriving and fattening upon the hard earnings of the industrious and peaceable part of the community."[6]

But such expenditures for defense also left the federal government decidedly short of cash. "Something must be done and done speedily," Secretary of the Navy William Jones wrote Pennsylvanian Alexander J. Dallas shortly before Dallas took over the Treasury Department, "or we shall have an opportunity of trying the experiment of maintaining an army and navy and carrying on a vigorous war without money."[7] By the time the new secretary made his first report to Congress later that fall, Dallas was estimating a revenue shortfall for the year of more than $12 million. Things only got bleaker looking ahead to 1815. Projecting expenditures of $56 million, including $15.5 million to service the national debt, Dallas calculated revenues—even with certain new taxes—to be only $15.1 million. Somehow, through a variety of loans and treasury notes, the government was going to have to raise more than $40 million to cover the difference.[8]

Many of Madison's Republicans in Congress were as aghast as were the Federalists. The story is told that Virginia Republican John W. Eppes of the House Ways and Means Committee read Dallas's report and threw it on the table with disgust. "Well, sir," he said to Federalist William Gaston of North Carolina, perhaps only half in jest, "will your party take the Government if we will give it up to them?" Gaston's response underscored the gravity of the underlying issue. "No, sir," he replied, "not unless you will give it to us as we gave it to you!"[9]

The Republicans remained in control and Congress passed a bill to rebuild Washington despite the entreaties of New York and Philadelphia. Another congressional measure authorized the purchase of Thomas Jefferson's personal library of some ten thousand books. These volumes would begin to replace the original Library of Congress that now lay in ashes. What could not be replaced, however, was President Madison's almost obsessive preoccupation with New England as the source of his problems. "You are not mistaken," Madison wrote long-time political ally Wilson Cary Nicholas, "in viewing the conduct of the Eastern states as the source of our greatest difficulties in carrying on the war; as it certainly is the greatest if not the sole inducement with the enemy to persevere in it."[10]

Great Britain had other reasons besides New England's discontent for continuing the war, but one thing was certain. In New England it was still very much Mr. Madison's war. The final straw was the administration's inability to protect its coasts proactively or at the very least reimburse the costs of the New England states doing so themselves. Why not keep federal taxes in the state to begin with, the Salem (Mass.) *Gazette* asked editorially, and while they were at it, make a separate peace with Great Britain by some measure of "Nantucket neutrality" extending to the entire state of Massachusetts? The newspaper went on to urge its neighboring states to join "a convention of alliance, amity, and commerce."[11]

It was not the first time that the New England states had heard such a call. The Louisiana Purchase, Alexander Hamilton's national banking system, and Jefferson's dread embargo had all fostered regional discontent. But now after two and a half years of an unpopular war that appeared to promise no early end, the Massachusetts legislature issued a call for a regional convention to weigh the differences and determine whether their resolution might call for so radical a solution as secession from the Union. Connecticut not only accepted the proposition, but also invited delegates to convene in Hartford. Massachusetts sent twelve delegates; Connecticut, seven; and Rhode Island, four. While the legislatures of New Hampshire and Vermont declined to sanction the convention officially, two delegates from a county in New Hampshire and one from a county in Vermont were seated, bringing the total to twenty-six.[12]

Were these twenty-six men really willing to dissolve the Union? "Through the whole Revolution," ex-Federalist John Adams, who had broken with his party over the necessity of war, wrote a friend, "the Tories sat on our skirts and were a dead weight obstructing and embarrassing all our efforts. They have now the entire dominion of the five states of New England."[13] But even among those who gathered at Hartford on December 15, 1814, that dominion was far more moderate than Adams or even Madison and his fellow Republicans feared.

The delegates to what came to be called the Hartford Convention were hardly revolutionaries—just New Englanders with a grudge. And some of them didn't even hold much of that. Their first act was to elect George Cabot of Massachusetts their presiding officer. Respected throughout New England, Cabot had made his money privateering during the Revolution, served a stint in the United States Senate, and at sixty-two explained his presence at the convention thusly: "We are going to keep you young hotheads from getting into mischief." One of Cabot's fellow Massachusetts delegates, Nathan Dane, was an equally elder statesman who years before had been instrumental in inserting the anti-slavery clause into the Ordinance of 1787 that established the Northwest Territory. Dane acquiesced in attending the Hartford

Convention only because "somebody must go to prevent mischief."[14]

Other moderates included Harrison Gray Otis of Massachusetts, who had worked to defeat Madison's reelection in 1812, Samuel Ward of Rhode Island, another of those who had marched against Quebec in 1775, and Roger Minot Sherman of Connecticut, whose father had helped to draft the Declaration of Independence. Indeed, the only avowed secessionist in the group appears to have been Timothy Bigelow, former speaker of the Massachusetts House of Representatives. Theodore Dwight, a columnist for the *Connecticut Mirror*, was chosen secretary of the convention, but even his acerbic lines penned against Republicans over the years had stopped short of urging secession.

What appears to have gotten this group of mostly moderates into trouble and alarmed not only their fellow New Englanders, but particularly President Madison and his administration, was their decision to hold their meetings in secret. Now, to be sure, there were ample precedents for this—the Constitutional Convention among them—but given the climate of the times, this secrecy only fueled wild public speculation about what treasonous whispers might be going on behind closed doors. While the *National Intelligencer* scoffed at the whole concept of closed meetings, saying that "they are valueless, and experience has convinced the nation that Congress has never kept a secret one week," the delegates to the Hartford Convention seem to have taken their pledge of secrecy to heart. The lack of even idle gossip emanating from the deliberations—what later news reporting would attribute to "a high-ranking administration official who spoke on condition of anonymity"—only fueled suspicions.[15]

High on the list of issues this closed-door group debated were the right of a state to preempt the federal collection of taxes within its borders for defense purposes and the right to nullify federal legislation that adversely affected its citizens. In this case, the issue was conscription. Could the federal government draft citizens into its armed forces and require state militia to serve under federal command, or could states nullify such laws to protect their citizens? Such debates, however, were taking place not only behind closed doors in Hartford, but also in Congress. A week before the Hartford delegates convened, a Federalist congressman from New Hampshire named Daniel Webster argued in the United States House of Representatives that states might in fact have the right to find certain "measures thus unconstitutional and illegal" to protect their citizens from "arbitrary power."[16]

Webster's comments that day in 1814 were only the opening round in a congressional debate over nullification that would rage for decades. Almost fifteen years later, Robert Y. Hayne of South Carolina stood on the floor of the United States Senate and delivered an impassioned plea for states' rights, including nullification of certain tariff provisions odious to the South. Ironically, when Hayne had finished, it was that same Daniel Webster, now a senator from Massachusetts, who rose to defend the Union as "one and inseparable." Nullification, Webster said then, could only lead to disunion or even civil war.

But if disunion was on the minds of some of the Hartford delegates, they left no formal record of it. In fact, Theodore Dwight went to great lengths some nineteen years later when writing his *History of the Hartford Convention* to stress that secession had not been on anyone's mind. (Of course, by that time, 1833, it was New England that stood firmly with the West against the nullification rumblings of the South.) What *was* on their minds—after the tax preemption and nullification debates—were seven proposed amendments to the United States Constitution. They were hardly subversive, nor for that matter even directly related to the conduct of the current war. Rather, they were indicative of New England's long-held disaffection for Thomas Jefferson and anything Virginian and the region's reluctance to accept its declining national influence as the country expanded westward. If the Hartford Convention had its way:

1. Representation and direct taxes would be apportioned only on the population of free persons—thus, by not counting African-American slaves as three-fifths of a person for such purposes, the South would lose seats in the House of Representatives.

2. New states could be admitted to the Union only with a two-thirds vote of both houses of Congress rather than a simple majority—that would slow the rush of new western states with their two senators each of equal footing.

3. Congress could not pass an embargo lasting more than sixty days—that left no doubt what New England thought of Mr. Jefferson's foreign policy.

4. Two-thirds of both houses had to consent to *any* interdiction of commerce with a foreign country—meaning even a sixty-day embargo could not be passed by only a simple majority.

5. Except in the case of invasion, two-thirds of both houses would be required to declare war—a provision that would have likely avoided the current conflict.

6. Only native-born citizens and those already naturalized could serve in Congress or hold civilian offices—this one was aimed at Swiss-born Albert Gallatin, Jefferson and Madison's secretary of the treasury.

7. Presidents could serve only one term, and no state could have two presidents in succession—so much for James Monroe and the Virginia line of succession to the presidency.[17]

The convention adjourned on January 5, 1815, and a report of its recommendations was published the following day in the *Connecticut Courant*. At first there was a collective sigh of relief—from Federalists and Republicans alike. The *New York Evening Post* opined that all would "read with vast satisfaction this masterly report, and rejoice to find [their] fears and alarms groundless."[18] But then, relieved that the Union was not being torn asunder, the two parties returned to the root of their differences. If President Madison meant to yoke the New England states with taxes and conscription, the *Boston Gazette* declared, he would need a swifter horse than he had ridden at Bladensburg. "He must be able to escape at a greater rate than forty miles a day, or the swift vengeance of New England will overtake the wretched miscreant in his flight."[19]

For their part, Republicans seemed more interested in focusing on what *might* have happened behind closed doors, rather than on the convention's written report. While "no mischief had been accomplished," Nathan Dane did say afterward that "if certain persons could have had their own way, and carried the measures which they proposed, I know not where we should have been."[20] And it was that possibility that Republicans rushed to condemn. Harrison Gray Otis became the principal defender of the gathering, and it cost him considerable public prestige. Only Massachusetts and Connecticut approved the recommended constitutional amendments, and eight states voted resolutions against them. The New York legislature espoused the Republican position, as well as that of growing public sentiment that the only effect of the amendments would be "to create dissentions among the different members of the union, to enfeeble the national government, and to tempt all nations to encroach upon our rights."[21]

In response to this growing suspicion of what *might* have happened at Hartford, the Federalists said little. Years later, historian Henry Adams succinctly summed up the resulting dilemma. "If any leading Federalist disapproved the convention's report, he left no record of the disapproval. In such a case, at such a moment, silence was acquiescence."[22]

And, of course, it didn't help the Federalists' image that some of them were predicting privately that "what might have been" was still going to occur. Timothy Pickering, an ardent Federalist not adverse to secession, wrote to John Lowell, who was of equal mind, that "with regard to the admission of new states into the Union, events with which the present moment is teeming may take away the subject itself." In other words, with certain British successes, there might not be half a continent available for expansion. "If the British succeed against New Orleans," Pickering went on, "and if they have tolerable leaders I see no reason to doubt of their success—I shall consider the Union as severed. This consequence I deem inevitable."[23]

Several years after Pickering's pronouncement—and after the fate of New Orleans had been resolved—Andrew Jackson wrote to James Monroe, who, despite New England's objections, had by then continued Virginia's hold on the presidency. Had Jackson "commanded the military department where the Hartford convention met," Old Hickory asserted, "if it had been the last act of my life I should have punished three principle [sic] leaders of the [Federalist] party . . . and am certain an independent court martial would have condemned them." Perhaps forgetting, or more likely attempting to blot out, his earlier association with Aaron Burr, Jackson went on to say that "these kind of men although called Federalists, are really monarchists, and traitors to the constituted government."[24]

In the end, the Hartford Convention was not so significant for what it did, as for what it failed to do—rally discontent from more than a handful of reactionaries. The moderates, it seems, just weren't biting. "I can affirm with confidence," Noah Webster wrote Daniel Webster in 1834, "that no body of men, of like number, ever convened in this country, has combined more talents, purer integrity, sounder patriotism and republican principles, or a more firm attachment to the constitution of the United States." That statement was made with the advantage of hindsight, but Noah Webster went on to characterize the taint the wild suspicions of "what might have been" had had on the entire Federalist Party. "The history of this convention," Webster asserted, "presents full proof that party spirit [Republican] may impose misrepresentations, upon a *whole people*, and mislead a great portion of them into *opinions directly contrary to facts*" [Webster's italics].[25]

The Federalist Party, long on the ropes, never recovered from the stigma of treason and disunion—no matter how unsupported by facts—that was attributed to it by the Hartford Convention. That is the real significance of the Hartford Convention. As historian Glenn Tucker opined, "The Federalist party died behind the closed doors at Hartford."[26]

Meanwhile, of course, the war went on.

CHRISTMAS IN GHENT

As another winter gripped the war-torn country, James Madison was in the throes of deep depression. His capital lay in ruins; voices in his own head if not in fact warned that half the country was plotting against him; and at sixty-three, he had been ill for much of the year. To make matters worse, dissent against his conduct of the war was not limited to the Federalist side of the aisle. "If we have another disastrous campaign in Canada," George Hay of Virginia had written to James Monroe in the spring of 1814, "the Republican cause is ruined, and Mr. Madison will go out covered with the scorn of one party, and the reproaches of the other."[1] Brown and Scott's actions at Lundy's Lane and Macomb's stand at Plattsburgh had not been disasters, but neither had they claimed any Canadian territory. The ensuing stalemate was an economic disaster. Somehow, the war that Madison had dreaded—the war that he had failed to prevent by more forceful use of his powers of presidential persuasion—had to be brought to a close.

Goodness knows Madison had tried to end the war almost from the day in June 1812 that Congress declared it. Scarcely a week after Madison signed Congress's declaration of war, Secretary of State Monroe advised the administration's chargé d'affaires in London, Jonathan Russell, to seek an early reconciliation with the British lion that required only two concessions:

repeal of the Orders in Council and cessation of impressment from American merchantmen. The first was easy, the second impossible.[2]

By the time Russell presented this olive branch to Great Britain's foreign minister, Lord Castlereagh, Parliament had already repealed the Orders in Council. That left impresssment. In the heated debate in the United States Senate on the day it voted for war, Senator James A. Bayard of Delaware, in making one last effort to postpone the matter, had made one unequivocal statement of fact. Stop impressment? "No war of any duration," the senator asserted, "will ever extort this concession—she [Great Britain] may as well fall with arms in her hands, as to seal quietly the bond of her ruin."[3]

Russell's offer to Castlereagh contained assurances that the United States Congress would quickly pass a law prohibiting the employment of British seamen on American vessels—thus giving Great Britain no reason to stop and search its vessels—but Castlereagh could hardly contain his disdain. He thought that the repeal of the Orders in Council should be more than enough to appease the Americans and expressed shock at their seeming intransgience on the impressment question.

"I cannot refrain on one single point from expressing my surprise," noted His Lordship, "that as a condition preliminary even to a suspension of hostilities, the Government of the United States should have thought fit to demand that the British Government should desist from its ancient and accustomed practice of impressing British seamen from the merchant ships of a foreign state, simply on the assurance that a law shall hereafter be passed to prohibit the employment of British seamen in the public or commercial service of that state." Moreover, Castlereagh's government simply could not "consent to suspend the exercise of a right upon which the naval strength of the empire mainly depends."[4]

And there the matter had rested for the remainder of 1812 and all of 1813. Both parties continued to fight a war that each claimed it didn't want, but that neither would end. During this time, Czar Alexander I of Russia offered to step into this impasse and mediate an end to hostilities. Having beaten Napoleon back from the gates of Moscow, Alexander was eager to reap the economic benefits that trade with both Great Britain and the United States—unhampered by naval warfare—would have in rebuilding his country. At the same time, the czar hoped to focus British attention on the final defeat of Napoleon and befriend the United States so that Britannia would not emerge from its global wars too supreme a power. (Indeed, Alexander's offer to mediate the War of 1812 was the first step in a Romanov policy of friendship with America—as a check on the British Empire—that was to continue through the American Civil War and culminate with Russia's sale of Alaska to the United States in 1867.)

Madison was all too glad to accept the Russian offer, and in March 1813 the president appointed James A. Bayard, Albert Gallatin, and the American minister to Russia, John Quincy Adams, to act as peace commissioners under the czar's mediation. The British government, however, declined the Russian offer, claiming that the matters in question—principally impressment— were internal to the governments of the belligerents and not matters of international concern. Nonetheless, by November 1813— with Napoleon momentarily defeated at Leipzig—Castlereagh offered to open direct peace negotiations with the United States. The United States formally accepted on January 5, 1814, and Madison quickly added Henry Clay and Jonathan Russell to the American commission.[5]

It is difficult to overstate the strength of the American delegation. Divergent though the five American peace commissioners were in their personalities, backgrounds, and individual views, collectively they were nonetheless arguably the most distinguished American diplomatic mission ever assembled. Their chairman, John Quincy Adams, had cut his diplomatic teeth serving as his father's secretary in Paris during the negotiations to end the Revolution. George

Washington sent the younger Adams to Holland and Portugal as chief of the American missions there, and as president, John Adams sent his son to Prussia in 1797. Elected to the Senate in 1803, John Quincy, like his father, also broke with the Federalist Party, and Madison appointed him minister to Russia in 1809. "If ever there was a just cause for war in the sight of almighty God," Adams had written from St. Petersburg at the war's outset, "this cause is on our side. We are fighting for the sailors' cause—the English cause is the press gang."[6]

No one had to ask Henry Clay what cause he was fighting for. Monroe and the rough-cut westerner were of equal minds: Canada, Canada, Canada. If the Madison administration had any fault, Clay believed it was in "betraying too great a solicitude" for peace, "an honorable peace [being] attainable only by an efficient war." Left to Clay, the nation's resources would have been marshaled and the war incessantly prosecuted until peace terms were dictated at Halifax or Quebec. That Clay would resign his speakership of the House of Representatives and accept Madison's appointment to the peace commission is a commentary on Clay's commitment to this goal as well as on the still minor power of the speakership. Doubtless, too, Clay calculated that such an assignment might bolster his chances to be named secretary of state in a future administration, that being but a stepping-stone to his ultimate goal of the presidency.[7]

Swiss-born Albert Gallatin was the nation's financial watchdog. After briefly serving in the Senate from Pennsylvania, he had become a leader of the Republican minority in the House of Representatives before Jefferson tapped him as secretary of the treasury in 1801. Madison retained him, of course, but Gallatin had the unpleasant task of both enforcing Jefferson's embargoes and managing their subsequent dire impacts on the treasury.

James A. Bayard of Delaware was a Federalist, but a moderate one, so much so that it had been his vote in 1801 that elected Jefferson president over Burr. A foe of the embargoes, he had done his best to avoid or at least delay the war but to no avail.

Bayard alone among the American delegates seems to have understood how intractable the British were on the issue of impressment.

Only Jonathan Russell among the commissioners was a lightweight, but this was more because of the company of the other four than his own experience. Appointed chargé d'affaires in Paris by Madison in 1810, Russell was sent to London the following year. Because he had been on the scene and because Madison proposed naming him ambassador to Sweden where the direct talks were initially scheduled, Russell was a logical and informed choice.

Federalist, Republican; westerner, easterner; war hawk, dove; there was something for everyone in this delegation.

As this assemblage made its way first to Göteborg, Sweden, and then Ghent in present-day Belgium, the British procrastinated. Napoleon was finished, and the legions that had vanquished him were en route to America. Perhaps they could wrest some territorial concessions or at the very least make Canada secure before hostilities ended. Thus, it was May 1814 before the British delegates were appointed and August 2 before they departed London for the short journey across the channel to Ghent.

Meanwhile, the Americans chafed at the delay. "They have kept us waiting nearly four months since the arrival of Mr. Clay and Mr. Russell in Europe, and their commissioners are not yet here to meet us," fumed John Quincy Adams in mid-July. "In the mean time they have sent to America formidable reinforcements . . . and I can imagine no other motive for their studied and long protracted delays to the commencement of the negotiation, than the intention of waiting for the effect of their forces upon our fears. Whatever they may do, I trust in God that they will find in our country a spirit adequate to every exigency. . . ."[8]

Compared with the American delegation, the British commissioners—when they finally arrived in Ghent—were second-tier. To begin with, Foreign Minister Castlereagh chose to send only three. Like the Americans, however, each had his particular constituency or expertise. William Adams was an admiralty lawyer, somewhat obscure to be sure, but guaranteed to know the finer points of maritime law ad nauseam. Admiral Lord Gambier was very much a desk admiral, but he could be counted on to know the utter importance of impressment to the Royal Navy. Henry Goulburn, at thirty the youngster of both delegations, was nonetheless well versed in North American matters as undersecretary for war and the colonies. If Canada was Clay's to take, it was Goulburn's to keep.[9]

But the fact was that these three British commissioners were not going to be so much as sneezing without direct instructions from London. Ghent's proximity to the British capital allowed foreign secretary Lord Castlereagh and secretary for war and the colonies Lord Bathurst to hold a tight rein on their representatives while they themselves were preoccupied with the far more pressing matter of carving up post-Napoleonic Europe at the Congress of Vienna. The Americans, by necessity, were on their own—communications with their government might take a couple of months instead of a couple of days.[10]

Monroe, of course, had provided the Americans with a steady stream of instructions. "On impressment," the secretary of state informed Clay before he sailed for Europe, "the sentiments of the President have undergone no change. This degrading practice must cease."[11] Three weeks later, on February 14, 1814, Monroe wrote again with the hope that Great Britain's recent victories on the continent might make it more inclined to exercise some degree of forbearance on impressment.[12] But John Quincy Adams feared just the opposite. Echoing sentiments Gallatin had already written to Monroe, Adams voiced his opinion that Great Britain's success in Europe "has undoubtedly made the continuance of the war with America, a purpose of policy with them, as much as it is a purpose of passion with their nation."[13] No wonder the British commissioners were dragging their feet getting to Ghent.

In June Gallatin reiterated this fear of delay to Monroe and told him point-blank that with no prospect of help from Europe, the best that the Americans could hope for was a draw—a return to the status quo and that which had existed in June 1812. "You may rest assured of the general hostile spirit of this nation [Great Britain]," Gallatin wrote, "and of its wish to inflict serious injury on the United States; that no assistance can be expected from Europe; and that no better terms will be obtained than the *status ante bellum.*"[14]

But even before Monroe received this, President Madison had come to acknowledge that the handwriting was on the wall. At a cabinet meeting on June 27, 1814, the question was asked, "Shall a treaty of peace, silent on the subject of impressment, be authorized?" With all replying in the affirmative, Monroe now instructed the commissioners that "you may omit any stipulation on the subject of impressment, if found indispensably necessary to terminate it. You will, of course, not recur to this expedient until all your efforts to adjust the controversy in a more satisfactory manner have failed."[15]

And so the poker playing began. Some say that the battles at the negotiating table—really an exchange of polite but firm letters—were fought as fiercely as were those on the battlefields. Indeed, as the full brunt of British power was brought to bear on the United States during the summer of 1814, the question was no longer "*What* can we get out of this?" but rather "*How* can we get out of this intact?" Each week that went by might bring news that Prevost was advancing on New York and that Washington's destruction was but the first of many similar fates. "I tremble indeed whenever I take up a late newspaper," Henry Clay wrote home to William Crawford in October. "Hope alone sustains me."[16]

While the Americans had been instructed to bluff on impressment as long as possible before folding, the British commissioners had been advised not even to discuss it. That didn't leave much to talk about on that subject. Instead, the British advanced a similar bluff position meant to take up time, but from which they were secretly willing to retreat. The subject was Indian rights, specifically the status of the Northwest nations that now lay decimated in the wake of Tecumseh's collapsed confederacy.

Castlereagh wanted to establish an Indian buffer state roughly between the Great Lakes and the Ohio River and use this both to protect the Canadian border from encroachment and to slow the American march across the continent. But in presenting this demand, the British commissioners almost overplayed their hand despite their proximity to Castlereagh's supervision. They announced that the British required the inclusion of Indian nations in the peace conference and the establishment of a definite boundary and security guarantees for their lands.

Canadian fur traders—who still held Mackinac Island and the Great Lakes northwest of Lake Erie—liked this idea. In fact, they had promoted a memorandum to the British commissioners that went so far as to suggest brashly that the eastern boundary of the proposed Indian buffer state should run from Erie, Pennsylvania, to Pittsburgh, and down the Ohio River. So much for the state of Ohio! That one made Henry Clay shudder. In fact, even a more modest proposal that the boundary should be the Wabash still had the United States giving up territory previously wrestled away from the Indians by William Henry Harrison after Tippecanoe. The Americans were as indignant over this issue as the British were over impressment.

Adams told the British a flat no. Henry Goulburn could only report to Lord Bathurst that "till I came here I had no idea of the fixed determination which there is in the heart of every American to extirpate [sic] the Indians and appropriate Their territory." Castlereagh, for his part, was appraised of the American reaction as he passed through Ghent en route to Vienna. Castlereagh fretted that his minions had been a little too zealous in presenting the British case and that Indian representation at the peace table and cession of vast chunks of claimed land might be just the fuel to rally American support for the war. So the British backpedaled a little and the exchange of letters continued.[17]

• • •

And what about Canada? In an 1813 letter of instructions to Adams when Russian mediation was still a possibility, Monroe had persistently, if not naively, declared: "It may be worthwhile to bring to view the advantages to both countries [Great Britain and the United States] which is promised, by a transfer of the upper parts and even the whole of Canada to the United States."[18] Now the British demanded territory in northern Maine to link Quebec and Nova Scotia—Maine east of the Penobscot River was still in British hands—demilitarization of the Great Lakes; and navigation rights on the Mississippi in exchange for fishing concessions in the Grand Banks. All this must be done, the British contended, to renounce once and for all America's obsession with conquering Canada.

Conquer Canada? When the Americans denied with straight faces that this had ever been their government's intent, the British dusted off General William Hull's pompous annexation proclamation to the people of Upper Canada upon his abortive foray from Detroit in 1812. Oh, that. Not really government-sanctioned, was the reply.[19]

Then came the news that Washington lay in ashes. Henry Goulburn sent Henry Clay a bundle of newspapers containing the reports and enclosed a sarcastic note. Later Goulburn crowed to Castlereagh: "We owed the acceptance of our article respecting the Indians to the capture of Washington; and if we had either burnt Baltimore or held Plattsburgh, I believe we should have had peace on the terms you have sent to us in a month at latest."[20] But, of course, Baltimore still stood, and Prevost was long gone from Plattsburgh. Cochrane might well capture New Orleans and force the issue of navigation on the Mississippi, but the doctrine of *uti possidetis*—that each side could keep whatever it possessed by conquest at war's end—was unpalatable to the Americans and looking less and less advantageous to the British.

Into the fray stepped the Duke of Wellington. When Castlereagh and the British cabinet asked the Iron Duke his advice, Wellington was candidly blunt. Given Prevost's withdrawal and the stalemate at Niagara, the British had no territorial claims, and the only way they might yet exert some was to establish control of the Great Lakes—something that Yeo had consistently failed to do. Great Britain had not won a foot of American soil and was fortunate in having lost none of its own. *Status ante bellum*. Take it and run, Wellington advised.[21]

So the peace commissioners sat down for one more round. Hostilities would end and "all territory, places, and possessions whatsoever, taken by either party from the other during the war" would be "restored without delay." The messy issues of certain minor boundaries and locations, such as several islands in Passamaquoddy Bay in eastern Maine, would be resolved by future joint commissions. There would be no concessions on either side regarding maritime rights. Impressment was not mentioned. And the Indians? Article IX of what became the Treaty of Ghent simply called on both belligerents to end all hostilities with the Indians "with whom they may be at war" and to restore them to "all the possessions, rights and privileges which they may have enjoyed, or been entitled to in 1811."[22]

This last was an empty promise on both sides. Great Britain could not restore Tecumseh's confederacy any more than the Americans were willing to repeal Harrison's land grabs after Tippecanoe or Jackson's after Horseshoe Bend. On its face, the Treaty of Ghent promised the status quo, but if the two main protagonists were to be made whole, the real losers were the Native Americans who had found themselves in their way. After 1814, the Northwest Territory and the states that would quickly join Ohio were indisputable. So, too, was the rapid admission of new states that would soon march across the Gulf Coast to join Louisiana. Diplomats may decree, but time does not stand still. History is always irrevocable. And so it was on the borders of the United States, Treaty of Ghent or not.

There were minor points of clarification to be made by both sides, a correction here, a change of word there. Finally, at six o'clock on the evening of December 24, 1814, the eight peace commissioners gathered around a long table and signed a "Treaty of Peace and Amity between His Britannic Majesty and the United States of America." As Jonathan Russell noted, "the Great Congress at Vienna" had overshadowed the "little Congress at Ghent." But if their governments should choose to act on it, peace was at hand.[23]

"I cannot close the record of the day," John Quincy Adams confided to his diary that Christmas Eve, "without a humble offering of gratitude to God for the conclusion to which it has pleased him to bring the negotiations for peace at this place." Now it remained for the treaty to be ratified by the respective governments.[24]

Across the Atlantic and at least six weeks away by favorable sail, the United States still wrung its hands in despair, ignorant of the action at Ghent. What was to be its fate? No less a public figure than Thomas Jefferson expressed great anxiety for where the British fleet might next appear and what impact its movements might have on the ratification of any peace treaty. Might not a last-minute victor seek to adjust its terms to his own benefit? Jefferson was particularly worried about New Orleans, convinced that the city would fall. If that happened, Jefferson expected the British to hold it indefinitely.[25]

ALONG THE
MIGHTY MISSISSIP'

With the possible exception of Tchaikovsky's *1812 Overture* celebrating Napoleon's defeat at the gates of Moscow, name one song closely identified with the War of 1812. If you are a baby boomer—or the offspring of one who ever endured a road trip with your parents—likely you know the answer. Almost a century and a half after the historic event, country-and-western singer Johnny Horton set folks to tapping their toes to the strains of his account of the Battle of New Orleans. "In eighteen-fourteen we took a little trip," sang Johnny, "along with Colonel Jackson down the mighty Mississipp' / we took a little bacon and we took a little beans / and we caught the bloody British near the town of New Orleans."[1]

Truth be told, of course, Andrew Jackson's "little trip" in 1814, was not *down* the mighty Mississippi, but rather via a circuitous loop through Alabama and along the Gulf Coast. And as for catching the British in New Orleans, Jackson was still expecting their landing at Mobile less than a month before they disembarked just east of the Crescent City.

It will be remembered that in August 1814, Jackson was at Fort Jackson in central Alabama, putting the finishing touches on his treaty with the Creek. At the same time, Vice Admiral Sir Alexander Cochrane was being given wide discretion to direct the third prong of Great Britain's grand 1814 strategy. Cochrane's first thrust southward was in mid-September, when a British naval force tried to seize Fort Bowyer at the entrance to Mobile Bay— even as the bulk of his fleet was still engaged in Chesapeake Bay. The Americans repulsed the attack, but it confirmed Jackson's long-held suspicions that the British were actively courting Spanish and Indian allies in the Floridas in preparation for a major invasion of the Gulf Coast.

The previous spring, General James Wilkinson had halted at the Perdido River after his capture of Mobile. Since then, Jackson had been warned not to antagonize Spain by invading that portion of West Florida that still remained in Spanish hands east of the river. Secretary of War James Monroe's most recent admonishment on this point was somewhere in transit when Jackson nonetheless crossed the Perdido and arrived at Pensacola with four thousand men early in November 1814. The Spanish quickly surrendered the little town and its surrounding forts, although British ships were able to evacuate some Spanish and British troops and blow up Fort Barrancas and Fort Santa Rosa at the mouth of the harbor before sailing away.

Unaware that Jackson had captured Pensacola, Monroe wrote a second letter to him expressing the hope that Old Hickory would take no actions that might annoy the Spanish. It was now too late for that, but Monroe's letters may well have been meant only as diplomatic window dressing. From a military standpoint, control of the fine harbor at Pensacola was highly advantageous to thwarting any British invasion of the eastern Gulf Coast, and both Jackson and Monroe knew it. Indeed, Admiral Cochrane himself confessed as much after Jackson captured the town.

Jackson felt just as strongly about Mobile. He saw it as the gateway both to the country along the Alabama River that he had just wrestled from the Creek and to a land route over which to attack New Orleans. Part of his concern stemmed from continuing reports that the British expected to rally a huge confederation of Indian tribes and black slaves to their banner and then proceed to march across the entire South.

But New Orleans at the mouth of the Mississippi was critically important, too. No one questioned that. Not only was it the gateway to a river system that ultimately led to Canada, but also on its docks and in its warehouses sat innumerable goods waiting for export—kept there by the British blockade. By one account, there was cotton alone worth 3.5 million pounds (about $15 million, or what the United States had recently paid France for the entire Louisiana Territory). It was a veritable treasure ripe for the picking. Thus, Jackson's dilemma was how to defend all these strategic points against a British amphibious invasion that might come anywhere along the Gulf Coast from New Orleans to Pensacola. Given the abysmal lack of adequate roads, it would take Jackson weeks to move forces between these points, while a British fleet could sail between them in a matter of hours.

While still in Pensacola, Jackson received intelligence that the British target was New Orleans. He remained somewhat skeptical, but could ill afford to ignore these reports. Leaving a large garrison at Mobile under the command of General James Winchester—the same Winchester whose troops had met with defeat at the River Raisin—Jackson finally hurried to New Orleans late in November 1814.[2]

But Cochrane's plan to attack New Orleans was far easier made than executed. To be sure, there was a plethora of routes into the city. That fact alone was enough to give defenders fits. From which quarter would the attack come? But none of the routes was without its disadvantages. First, there was the main channel of the Mississippi River. It led to the city but required sailing ships to negotiate eighty miles of the river's twisting meanders and strong currents. Fort St. Philip was an American outpost thirty miles above the mouth, and Fort St. Leon at the English Turn just below the city was another.

There were three routes from Lake Borgne. One led up the Chef Menteur road to the Plain of Gentilly east of the city. A second led up Bayou Bienvenu to near the Mississippi about ten miles below the city, and the third led through the watery maze of the Rigolets into Lake Pontchartrain. From the lake, Bayou St. Jean led straight south to the city's back door. The drawback to all the Lake Borgne routes was that shallow-draft vessels were required to navigate its waters. The main British fleet would have to be left at anchor some eighty miles away. There were possible routes up Bayou La Fourche to reach the Mississippi upstream from New Orleans and up Bayou Terre Aux Boeufs to reach the river at the English Turn from the east, but both of these routes were narrow and more easily defended. And then there was Barataria Bay.[3]

Few aspects of the Battle of New Orleans are wrapped in denser mists of legend and intrigue than the story of Jean Lafitte and his fellow Baratarians. Pirate, scoundrel, gentleman, patriot. He has been called all of those and more. What truth there once was has long since been distorted by sensationalist novelists and more than a few motion pictures. Perhaps the most famous movie, the 1958 version of *The Buccaneer*, starred Charlton Heston as a backwoods Andy Jackson and Yul Brynner as a dashing Jean Lafitte. It is actually a rather accurate portrayal of the key events leading up to the battle—save for the subplot of Lafitte's romance with Governor Claiborne's daughter.

By one account, Jean Lafitte was born in Port-au-Prince, Haiti, in 1782, his family having been driven there from Spain. He had at least two older brothers, Pierre and Alexandre, the latter of whom adopted the alias of Dominique You. Dominique was well-established as a privateer and had acquired a reputation as an able marksman with a cannon. Pierre and Jean learned the seafaring trade from their relative, Renato Beluche, and like him became smugglers and privateers, preying on ships of all countries and in time establishing bases of operations in and around New Orleans. Such activities put the brothers on a collision course with the newly

arrived American government after 1803, and tensions only escalated after Louisiana became a state in 1812. One reported episode
has an exasperated Governor William C. Claiborne offering a
reward of $500 for Jean Lafitte, only to have the pirate offer a
reward of ten-fold that for the governor.[4]

Suffice to say that from years of smuggling, Jean Lafitte and
his band knew the southern approaches to New Orleans like the
backs of their hands—particularly their stronghold at Barataria
Bay south of the city. In addition, Lafitte and his men and ships
would be a valuable resource under whichever flag they might
choose to fight. The British made the first overture, and on
September 3, 1814, the brig HMS *Sophia* sailed into Barataria and
sought out Lafitte. Royal Navy captain Nicholas Lockyer was
persuasive and encouraged Lafitte to support the British efforts.
Lafitte was noncommittal, even stalling the British while he sent
word to the Americans of their visit. Offering his services to the
Americans instead, Lafitte was repaid for his efforts two weeks
later by an attack from American gunboats commanded by
Commodore Daniel T. Patterson, which laid waste to Lafitte's
main base of operations on Grande Terre.[5]

To add insult to Lafitte's injury, Andrew Jackson made it quite
clear to Governor Claiborne long before he arrived in New
Orleans that he wanted no part of Lafitte and his "hellish ban-
ditti" and that they should be "arrested and detained, until further
advice."[6] Still, Lafitte turned down the British offer despite its
sweeteners of gold and a commission in the Royal Navy and
chose to bide his time.

So, whatever else they could count on, the British could not
count on the services of the intrepid Lafitte. But they had more
complications than that. Admiral Cochrane's grand plan to capture New Orleans and the booty that went with it was almost
immediately muddied by two nagging problems. The first was a
lack of secrecy. "On my arrival at Jamaica, I found to my very
great astonishment," Cochrane wrote the Admiralty, "the intention of sending an expedition against New Orleans and Louisiana

which I had taken the utmost precautions to keep profoundly secret, publicly known throughout Port Royal and Kingston. . . ." This leak, Cochrane continued, had caused his adversary, Jackson, "to relinquish his intentions" to remain at Pensacola with an army of three thousand men and instead "proceed immediately for New Orleans."[7] Consequently, Cochrane did not dally in Jamaica. The admiral sailed with the bulk of his fleet on November 26, 1814, determined to beat Jackson to New Orleans.

In doing so, Cochrane soon discovered his second problem. Despite his orders to the contrary, there was a considerable shortage of shallow-draft vessels among the fleet. Unless the British fleet sailed directly up the Mississippi—an approach made problematic both by the vagaries of the river's course and current and the American fortifications—Cochrane would have to attack via one of the Lake Borgne routes. Had there been an adequate number of shallow-draft vessels, Cochrane appears to have favored passing through the Rigolets from Lake Borgne into Lake Pontchartrain, effectively arriving at New Orleans's weak backside. But now with only a limited supply of shallow-draft boats, no matter where he was going, Cochrane would be forced to ferry troops and supplies across Lake Borgne from an anchorage seaward of Cat Island at the lake's entrance.

What smaller ships Cochrane had at his disposal quickly bested five American gunboats on Lake Borgne and then in concert with a collection of oar-powered flatboats began to ferry troops and supplies between the anchored fleet and boggy Pea Island on the north shore of Lake Borgne. Any thoughts of advancing all the way into Lake Pontchartrain were quickly discouraged both by exaggerated reports of American defenses at the Rigolets and the fact that this route would have more than doubled the distance his men had to row between fleet and beachhead. Major General John Keane, an Irishman with distinguished service on the continent, led the assault brigade from Pea Island on December 22, 1814, as it rowed across the remainder of Lake Borgne and landed at the mouth of Bayou Bienvenu unopposed.

A distance of roughly five miles separated the bayou from the Mississippi River about a dozen miles downstream from New Orleans. Moving inland up the bayou and then striking toward the river, Keane's advance guard, led by Colonel William Thornton's stalwart Eighty-fifth Regiment, captured the Villeré plantation before noon on December 23 and made it their head-quarters. General Jacques Villeré (the local militia seems to have had a preponderance of generals at the time) was captured, but his son, Major Gabriel Villeré, managed to escape by jumping out a window. Gabriel hastened to New Orleans to warn Jackson that the British had in fact landed and done so in force.

Ahead of Keane, the route into the city looked open and promising. Some fishermen on Lake Borgne told the British that Jackson had only two thousand men with him. Keane's continued rapid advance by a disciplined brigade of like number might have burst the city wide open before Christmas. But there were also wild reports that Jackson had ten times that number with him. Keane elected to be cautious. His caution gave Jackson time to rally more troops.

In fact, Jackson probably had about four thousand troops at his disposal—the exact number being unknown even to him because of the fluid nature of militia arriving and departing and the many mixed units in his command. But Jackson correctly sensed that whatever troops Keane had put ashore, they were but the advance guard of the entire British army, and he determined to strike a blow before they could be reinforced. Posting General William Carroll's Tennessee brigade along the Gentilly Road to guard against the attack he still feared would come directly from Lake Pontchartrain, Jackson marched several thousand men down the east bank of the Mississippi straight at the British encampment on the Villeré plantation. "By the eternal," legend has Old Hickory exclaiming, "they shall not sleep on our soil! Gentlemen, the British are below, we must fight them to-night."[8]

Meanwhile, Jackson directed the navy's Commodore Patterson, who had recently paid a visit to Lafitte's lair on Barataria, to move the fourteen-gun schooner *Carolina* downriver to bombard the British position. When the *Carolina*'s guns opened up at about 7:30 on the evening of December 23, Jackson's troops launched their land attack by the light of a full moon. Initially caught unawares, Keane's troops were huddled around large campfires trying to eat and stay warm. This was not the balmy weather they had been promised when they left Jamaica.

The *Carolina*'s cannons caused confusion, and Jackson's right-hand man, General John Coffee, led his brigade on a sweep to try to turn the British right flank and pin Keane's entire force against the river. But the British formed up and counterattacked, aided by a ground fog that now caused Coffee's brigade to become separated from the other American units. By 9:00 P.M. the British line had held, and Jackson ordered a withdrawal to a defensive position along the Rodriguez Canal between the Macarty and Chalmette plantations. The first phase of the Battle of New Orleans was over.[9]

Two days later, just before noon Christmas Day, both sides were busy improving fortifications along their respective lines, when British cannon boomed a welcome to Lieutenant General Sir Edward Pakenham, General Ross's appointed successor and the brother-in-law of the Duke of Wellington. Pakenham's troops knew of his bravery and success at the side of the Iron Duke, and they expected no less of him here. There were even rumors that Pakenham carried with him a commission to serve as governor of Louisiana once the ragtag Americans were routed.

But Sir Edward did not like what he found. The reinforcements that had arrived with him were pouring into "a sort of *cul de sac*" around the Villeré plantation. The entire British position was wedged uncomfortably between the river on its left and a morass of wild swamps on its right. Pakenham was also decidedly not pleased with Keane's failure to advance immediately upon New Orleans. "I regret the defeat of our forces due to the error made on the 23rd of December," Pakenham huffed. "Our troops should have advanced to New Orleans immediately on taking Villeré's plantation."

But Admiral Cochrane took strong exception to this characterization. "We were not defeated," the admiral maintained, "and there is nothing wrong with our position." Then came his taunt: "If the army shrinks from the attack here," Cochrane brashly told Pakenham, "I will bring up my sailors and marines from the fleet . . . and march into the city." Perhaps, as some historians have suggested, Cochrane was indeed counting the loot to be had in that event.[10]

The next morning, Pakenham rode forward to survey the American line for himself. What he saw soothed him somewhat. There were a few horsemen galloping around "in a most unmilitary fashion" and firing now and then at the British pickets. They gave "the appearance of snipe and rabbit hunters beating the bushes for game" and hardly appeared of the caliber to trouble the might of the British Empire. In fact, the British rank and file, resplendent in their multicolored uniforms, called their American opponents "dirty shirts." Pakenham seems to have concurred in this opinion and after surveying the field before him determined that he would stand firm and fight from here.[11]

Jackson, of course, was equally determined to hold his line on the Rodriguez Canal and keep the British from advancing any closer to New Orleans. In reporting the night engagement of December 23, Jackson noted that "since then both armies have remained near the battle-ground, making preparations for something more decisive."[12] On Jackson's side, these preparations included the erection of a stout rampart running along the canal from the river some twelve hundred yards into a large cypress swamp. The breastworks varied in height and thickness but afforded a strong position, particularly when the canal in front was flooded and turned into a moat. Artillery pieces were positioned in key locations along the line and set on wood planking laid atop cotton bales to prevent the heavy ordnance from sinking into the mud. When one New Orleans merchant complained that his expensive cotton was being used for such purposes instead of an inferior grade, one of Jackson's aides replied, "Well, Mr. Nolte, if this is your cotton, you, at least, will not think it any hardship to defend it."[13]

Jackson also took precautions on the west bank of the river. Commodore Patterson took cannon off the sloop *Louisiana* and positioned this marine battery so that its guns could deliver a raking cross fire against the field in front of the American rampart on the east bank. Three hundred yards in front of Patterson's marine battery, General David Morgan established the west bank's major line of defense. It was manned with a collection of regulars, militia, and newly arrived volunteers, including four hundred Kentuckians who reached the front poorly armed and utterly exhausted. Two six-pounders and one twelve-pounder strengthened this line and were also able to fire across the river. Both Patterson's and Morgan's positions became even more important to guarding the river itself after the schooner *Carolina* was blown up during a fierce duel with British artillery. Pakenham, too, quickly recognized the importance of these west bank positions and their artillery pieces. If they could be captured, their raking firing could also be directed against the American rampart and not the open field in front of it.

Consequently, having expressed his disdain for the slowness of the British advance, Pakenham devised a plan to correct the situation. Relying on arriving reinforcements, Sir Edward proposed a simultaneous, two-prong attack along both banks of the river. While part of his command attacked Jackson's line along the Rodriguez Canal directly, a force of at least fourteen hundred men would cross the river and attack the sparsely defended artillery positions of Commodore Patterson and General Morgan. Once these positions were taken, their guns would be turned on the American line across the river. Properly executed, it was just the sort of tactical plan that would rank Sir Edward with his storied brother-in-law.

But how to get a major force across the river? The despised flatboats and barges would have to be brought to the river from

Lake Borgne and Bayou Bienvenu. Once again, Admiral Cochrane was assertive in his advice. Dig a canal, he said. The distance from the upper bayou to the river was about two miles, and the expedient answer would have been simply to roll the wooden boats that distance on logs. It would have been back-breaking work over soggy ground, but the British had already used a similar tactic to get their artillery pieces into place. But the admiral remained insistent. No, that wouldn't do. Instead, Cochrane demanded the maritime solution and ordered that a canal be dug from the bayou to the river.

While work on the canal was progressing, Pakenham received reinforcements of two sparkling regiments totaling some seventeen hundred men. These were the Seventh Fusiliers, resplendent in red coats with blue facings, white tufts in their caps, and light blue knapsacks, and the Forty-third Light Infantry, uniformed in red coats with white facings, green tufts, and black knapsacks. Much more than pretty, they were possibly the best-trained combat veterans in the world. The regimental drums of the Fusiliers and the bugles of the Forty-third had struck fear into Napoleon's legions in Spain and the south of France, and they were ready for anything. Their commander was Major General John Lambert, an Englishman who had also served under Wellington and who hoped that this service in the wilds of America would add only more laurels to his personal reputation and those of his regiments.[14]

Determined to execute his plan, Pakenham now gave assignments to his four brigades. To Colonel William Thornton, who had led the way at Bladensburg and who was perhaps one of his ablest lieutenants, Pakenham delegated the crucial task of getting across the Mississippi and seizing the American positions on the west bank. Thornton's success, however, depended on Cochrane's canal, and the admiral's textbook solution turned into a real quagmire. Trying to dig through the rain-saturated Mississippi mud was akin to spooning a trough through a bowl of soupy tapioca pudding. The walls of the would-be canal quickly collapsed and clogged what passageway could be dug. The result was that only about a third of the boats required to transport Thornton's force made it through, and then only after interminable delays. Thus, with boats for only five hundred men instead of the original fourteen hundred, Thornton finally departed the east bank eight hours late in the wee hours of January 8, 1815.[15]

On the east bank, the British attack would be led by Major General Samuel Gibbs and his brigade of twenty-two hundred men against Jackson's left and General Keane and his brigade, including Keane's own regiment, the Ninety-third Highlanders. General Lambert's newly arrived regiments were to be held in reserve. A central part of Gibbs's success depended on the role of the Forty-fourth, an Irish regiment. Its task was to storm the canal ahead of Gibbs's advance and throw bundles of sugarcane (fascines) into the Rodriguez Canal to form a bridge, after which scaling ladders could be placed against the American rampart. When the Forty-fourth's commanding officer, Lieutenant Colonel Thomas Mullens—one who had bought his commission rather than earned it—heard his orders, he is reported to have said, "My regiment has been ordered to execution. Their dead bodies are to be used as a bridge for the rest of the army to march over." A company of West Indian troops was to provide a similar service at the head of Keane's brigade. Both assignments seemed less grim to Pakenham because he surmised that the ragtag Americans would flee at the sight of the oncoming British might long before his troops reached the canal.[16]

No one on either side got much sleep on the night of January 7, 1815. Men in both armies, from Jackson and Pakenham down to the greenest recruit, sensed that "something more decisive" was about to happen. Truth be told, few had been sleeping very well for over a month. It was hard to do so when one's woolen garments—whether the finely tailored threads of the British regulars or the homespun of Jackson's militia—were continuously wet from daily rains and soggy ground. The air was cold and damp, typical Louisiana bayou weather for January. At night, the chill in the air struck daggers into the hardiest and reduced many to shivering uncontrollably.

Over on the west bank, Commodore Patterson assumed his nightly position above the river and stared into the darkness. From the east bank this night came new sounds of commotion. These were Colonel Thornton's troops preparing to disembark, however belatedly. But in the darkness, the din sounded as if half the British army was on the move, and Patterson was suddenly seized with the fear that the British intended to outflank Jackson by concentrating their attack up the west bank. The commodore hurriedly dispatched an aide to cross the river and ask Jackson for immediate reinforcements.[17]

Jackson was trying to get some sleep on a couch in the parlor of the Macarty house. He was on his feet as Patterson's aide was shown into the room. Patterson was mistaken, Jackson said. "The main attack will be on this side, and I have no men to spare. He must maintain his position at all hazards." While the aide hurried back across the river, Jackson roused his staff as they slept fully dressed with their swords and side arms by their sides. "Gentlemen," Old Hickory opined, "we have slept enough. Rise. The enemy will be upon us in a few minutes. I must go and see Coffee."[18]

Coffee. Jackson's stalwart friend, John Coffee. He had always been at Jackson's side whenever the going got tough, from the Benton gunfight, to the diversionary attack at Horseshoe Bend, and now in the misty cold on the banks of the Mississippi. Jackson knew that he could count on his friend, and the one thousand or so fellow Tennesseans and Choctaw in Coffee's command to anchor the left flank of his army, running along the Rodriguez Canal from the cane fields of the Chalmette plain into the cypress swamp. The British must not be allowed to turn this flank.

To Coffee's right on the line stood sixteen hundred more Tennessee troops under the command of General William Carroll. In the predawn hours of January 8, Jackson moved his reserve of one thousand Kentuckians under General John Adair into position about fifty yards behind Carroll's troops. Unbeknownst to him or any of the Americans, this was to be the very point that the British would soon concentrate their assault because an American deserter had previously reported it to be the weak link in Jackson's line.[19]

To Carroll's right—between his position and the river—the American right flank was held by an assortment of regulars, militia, and volunteers under the command of Colonel George Ross. These included Major Louis Daquin's battalion of about 150 Santo Domingo free men of color and a second battalion of free men of color commanded by Major Pierre Lacoste. They also included companies of Jean Lafitte's men.

Old Hickory, it seems, had had a change of heart about accepting help from pirates. Lafitte himself had visited the general and, much like William Weatherford after the Battle of Horseshoe Bend, Jackson had been impressed with his display of personal courage. Jackson was undoubtedly partly persuaded to accept Lafitte's assistance because Jackson now realized that he needed all the help he could muster against the British. But Jackson quickly warmed to Lafitte, so much so that the gentleman pirate was soon one of Old Hickory's trusted aides, advising on fortifications and furnishing seamen to man the guns of the *Carolina* and *Louisiana*. Now Jackson's right flank was not only bolstered by Lafitte's men, but pirate flints, powder, shot, and pistols were also in use along the line.[20]

To the rear of the right flank behind the Macarty house, Jackson held a company of fifty dragoons in reserve. Eight artillery batteries were spread along the canal, with battery number three being under the command of Dominique You. "I wish I had fifty such guns on this line, with five hundred such devils as those fellows behind them," legend has Jackson remarking of You and his Baratarians.[21] All told, Jackson had more than four thousand troops dug in along the Rodriguez Canal.

• • •

On the British side, Sir Edward Pakenham had not slept any better or longer than Jackson. Rising long before dawn, he rode from the Villeré house to the east bank of the river to determine the progress of Thornton's force. There was nothing to be seen of it—the mighty Mississippi had swept Thornton's reduced force far downstream of its proposed landing site. Surely, in the loneliness of those predawn hours, Pakenham must have known that the timing of his carefully planned attack was unraveling. No matter how determined Thornton might be, there was no humanly possible way that he could be ready to seize Morgan's position and guns in concert with the main British advance. Here was one last chance for Pakenham to ponder his fate and that of his army. But the die was cast, his troops primed for battle, and retreat in the face of these roughshod Americans unthinkable. Shortly after four o'clock, Pakenham turned from the river and remarked to an aide, "I will wait my own plans no longer."[22]

As Sir Edward rode back toward Gibbs's brigade, he passed the storied regiment of the Ninety-third Highlanders. Their commander, Colonel Robert Dale, had just learned of Thornton's delay in getting across the river. Turning to the regiment's physician, Dale took out his watch and a letter. "Give these to my wife," the colonel directed, "I shall die at the head of my regiment."[23]

Pakenham found General Gibbs in an agitated state. Mullens's Forty-fourth Irish Regiment was supposed to be in place at the head of his column with fascines and ladders, but they were nowhere in sight. Here was yet another delay and operational snafu that might have caused Pakenham to postpone the advance. But daylight was rapidly approaching and there was no time to waste. Without knowing for certain the disposition of the critical Forty-fourth, Pakenham turned and gave the order to fire the rocket signaling the attack. The greatest army in the world, the army that had bested Napoleon's legions, marched forward to the ruffle of drums, the blaring of bugles, and the mournful wail of the Highlanders' pipes.

The sun was indeed rising, but one could scarcely tell it amid the dark, dank fog that had rolled in from the river. To Pakenham, the fog was almost as good as darkness and would afford his troops cover as they advanced across the open field toward the American line. For some reason, Keane's brigade was momentarily confused by Pakenham's rocket signal and slow to join the advance of Gibbs's column. Finally, after several minutes that seemed interminably longer, the sound of British cannon on the right flank and the answering thunder of the Americans spurred Keane to advance. A third British column under Colonel Robert Rennie also moved forward along the river against Jackson's right flank.[24]

Onward the British came. But then the fog that Pakenham had counted on to save him from his delays began to lift under the warming rays of the rising sun. Within minutes the entire sweep of the plain at Chalmette—from cypress swamp to the river—was exposed to full view from the American lines. What an imposing sight! In colorful, well-appointed uniforms—damp and muddied but resplendent nonetheless—and with a sea of flags and banners flying, Pakenham's army came on, the bulk of it pointing squarely for General Carroll's place in the American line. Carroll's Tennessee men raised a cheer. Let them come! Behind them, Adair's Kentuckians—in the right place at the right time—raised a similar cry.

So across the plain at Chalmette the pride of the British army advanced. Three of Jackson's artillery batteries poured a devastating fire against the head of Gibbs's column. One American thirty-two-pounder, "filled to the very muzzle with musket balls," belched its load at point-blank range into the vanguard of the advancing column and swept "the center of the attacking force into eternity."[25]

But the British wave continued, and as it did, the American riflemen waited patiently behind their rampart. It would take more than fancy dress and flashy bayonets to make them run. Finally, when the British column was within two hundred yards of the Americans, Carroll shouted the command, "Fire!" and up and down the line it erupted with the rattle of musketry. Once again,

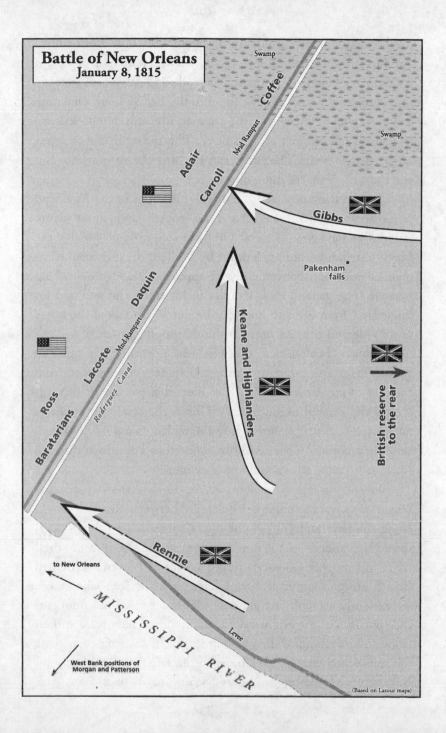

the battleground was suddenly obscured, this time from smoke "so thick that every thing seemed to be covered up in it."[26]

Safely behind their ramparts, Carroll's and Adair's troops continued to pour a deadly volley fire into the British lines. One rank would step up to the crest of the parapet, fire, and step back down to reload while another took its place. This meant that at the apex of the British attack, the American fire was almost nonstop. But still the British troops came on.

Close at hand now was the ditch at the foot of the American breastworks. Where was the Forty-fourth and those damn fascines and ladders? General Gibbs looked around. Finally, the Forty-fourth was coming, led not by Mullens, its disgraced commander who had conveniently disappeared, but by Pakenham himself. The general took a bullet in his right arm and had his horse shot from beneath him, but he quickly mounted the horse of an aide and pressed on into the developing chaos of Gibbs's front ranks. "For shame," the would-be governor of Louisiana shouted as troops threw down their knapsacks and recoiled from the fury of the American fire. "Recollect that you are British soldiers," the general urged them. "This is the road you ought to take," he exclaimed, pointing into the melee. "I am sorry to have to report to you," General Gibbs shouted to Pakenham through the din, "that the troops will not obey me."[27]

Now it was up to General Keane to save the day. Having begun his advance somewhat belatedly between Rennie's troops along the river and Gibbs's column, Keane ordered the Ninety-third Highlanders—some nine hundred strong—to move to their right and support Gibbs's faltering advance. "Never," recounted British ensign George Robert Gleig, "was any step taken more imprudently, or with less judgment."[28] For a few brief moments the approach of the Highlanders, one hundred men wide at their front and marching to the swirling strains of their pipers playing the regimental charge, gave hope to the faltering British troops. But then the American line roared again. Just as he had foreseen,

Colonel Robert Dale took grapeshot clean through his body and was killed instantly at the head of his regiment.

One hundred yards from the American line, the Highlanders paused. They were in a killing field. Pakenham suddenly realized that he was losing his army in the most humiliating of ways and ordered up Lambert's reserve, although it was stationed a good mile away. Then, removing his hat with his remaining good arm, Pakenham waved it at the Ninety-third and shouted "Hurrah! brave Highlanders!" Scarcely had these words escaped his mouth than a load of grapeshot exploded around him. Once more the horse Pakenham was riding was killed, and this time the general pitched to the ground with a wound in the thigh. Another round quickly struck him in the lower back, paralyzing him. Sir Edward was hastened to the rear, but died minutes later in the shade of one of the grand live oak trees.[29]

General Gibbs, too, fell with a mortal wound within twenty yards of the American ramparts. General Keane was wounded in the neck and thigh and taken from the field. Now there was no general officer left on the field to command the disintegrating British columns. General Lambert, who from his position well to the rear was suddenly the ranking British officer, later reported to superiors that this loss of general officers and the failure of the Forty-fourth to breach the ditch had "caused a wavering in the column, which in such a situation became irreparable; and as I advanced with the reserve, at about two hundred and fifty yards from the line, I had the mortification to observe the whole falling back upon me in the greatest of confusion."[30]

Lambert stood his ground, but by then it was over. As the smoke slowly drifted off the field, the view from both sides was shocking. "Did you ever see such scene?" asked one of Pakenham's dazed aides. "There is nothing left but the seventh and forty-third [Lambert's reserve]." The Americans looked out upon a field of mangled bodies and tattered banners. Of the three thousand troops in the main British advance, two-thirds lay dead or dying on the field. "It was like a sea of red, remembered one Kentucky militiaman, not because of the blood but because of the number of red coats lying side by side. One could walk on the bodies of the dead without touching the ground." Less than thirty minutes had elapsed since Pakenham's rocket signal.[31]

Some urged Jackson to order an advance and rout the remaining British force. He may have been tempted, but this was not Horseshoe Bend. Besides, despite the victory before him, there was trouble on the west bank. Having been swept four miles downstream from Morgan's position, Colonel Thornton nonetheless made admirable time marching at the double-quick to engage. He was well aware by now that Pakenham was fighting fiercely on the east bank. Assembling his troops some seven hundred yards in front of Morgan's forward position, Thornton ordered a feint toward the strongest American position along the river and then concentrated his attack to turn the American right flank. This was tenuously formed and manned by a detachment of Kentuckians commanded by Colonel John Davis. As Thornton's troops swept up to their front, the Kentuckians gave way and fled. Morgan rode up to Davis and exhorted him to halt his men, but Davis "replied that it was impossible." "Sir," Morgan retorted, "I have not seen you try."[32]

Morgan made an effort on his own to stop the flight, but by then Thornton had pressed the attack all along the American line so successfully that Morgan's artillery battery was forced to spike its cannon and pitch them into the river. Now all that stood between Thornton and New Orleans on the west bank was Patterson's marine battery that had been bombarding Colonel Rennie's advance along the east bank. Seeing Morgan's line disintegrate, Patterson turned his guns from the river to face Thornton's advancing troops. But Patterson's line was soon also breached and he, too, was forced to spike his guns and retire toward the city.[33]

Watching with his telescope from the opposite bank, Old Hickory was furious. His smashing victory might yet be undone. Hastily he dispatched four hundred men across the river to rally

Morgan's fleeing troops. "The Kentucky reinforcements, in whom so much reliance had been placed," Jackson later reported, "ingloriously fled—drawing after them, by their example, the remainder of the forces."[34]

Thornton quickly occupied Morgan's deserted position and, not knowing that his commander was dead, sent a message to Pakenham announcing his success. Accompanying it was a note from his artillery commander listing the captured American field-pieces and noting that one howitzer was inscribed, "Taken at the surrender of Yorktown, 1781." Had the Americans not spiked their guns, the British might easily have turned them on Jackson's line across the river and created havoc if not erased the holocaust that had already befallen their comrades. But that was not to be.[35]

Lambert quickly dispatched his artillery chief to survey Thornton's position. When he reported that it could not be held by fewer than two thousand men, Lambert ordered a retreat. All of Thornton's hard-won ground was given back. Had Thornton gotten off on schedule with his original force and reached the guns in time to turn them on the Americans across the river as Pakenham advanced, it might have been a different story. Looking for anyone to blame, the British hanged the American deserter whose information about the weakness of Carroll's position had led Gibbs's main thrust into the strongest part of the American line. He was branded a spy sent to spill misinformation.[36]

As so often was the case during this war, it was difficult to compile an accurate list of casualties in the aftermath of this horrific encounter. Jackson estimated British losses at 400 killed, 1,400 wounded, and 500 taken prisoner, while reporting his own as only 7 killed and 6 wounded. British estimates of their losses were only slightly lower: 291 killed, 1,262 wounded, and 484 captured. Of course, the British losses included not only their commander in chief, but General Gibbs as well. "Such a disproportion in loss, when we consider the number and the kind of troops engaged, must, I know," Jackson wrote the secretary of war, "excite astonishment." By anyone's count, it was a smashing American victory and a devastating British defeat.[37]

But what would the British do now? Bloodied they were, but acting commander Lambert still had the cream of his reserve, some seventeen hundred men of the sparkling regiments of the Seventh Fusiliers and the Forty-third Light Infantry, largely intact. And even as Pakenham had been slugging it out on land, Admiral Cochrane was attempting to make good on his boast that the Royal Navy alone could capture New Orleans by sailing directly up the Mississippi. For nine days after the main battle, Royal Navy forces attempted to subdue Fort St. Philip thirty miles from the river's mouth. When the Americans held firm, both Cochrane and Lambert decided that it was time to withdraw.

Lambert's troops hacked out a swampy road—about as successfully as Cochrane's infamous canal—to get his wounded back to Lake Borgne. "Last night, at twelve o'clock," Jackson reported on January 19, 1815, "the enemy precipitately decamped and returned to his boats, leaving behind him, under medical attendance, eighty of his wounded including two officers, fourteen pieces of his artillery, and a quantity of shot, having destroyed much of his powder."[38] Governor Claiborne sent letters to Jackson and his officers, profusely thanking them for their spirited defense of the city. Jackson himself would brag that "with vigilance I have defeated this boasted army of Lord Wellington."[39]

But the British were far from finished. Still blissfully ignorant of the Treaty of Ghent, Cochrane sailed east and once again attacked Fort Bowyer. This time it fell, and the British tightened a noose around Mobile. Just when it looked as if General James Winchester would be forced to surrender yet again to the British, the news from Ghent finally arrived and the British assault on the city was abandoned. No doubt swallowing a great deal of pride at the effort, Admiral Cochrane nonetheless wrote Jackson with the news. Sending Jackson a copy of a bulletin he had received from Jamaica, the admiral who had once covetously eyed the treasures of New Orleans was forced to "offer you my sincere congratulations."[40]

Great Britain would soon be occupied again on the continent of Europe, but its defeat at the hands of the rabble that had "stood beside our cotton bales and didn't say a thing"[41] would remain a thorn in British pride well beyond the zenith of Victoria's empire. It was, according to one British newspaper, "a real surprise . . . that they [the British army] ran into such strength." Why the British had expected that the melting pot of Louisiana with its strong French heritage should welcome them as liberators is difficult to say. But Louisiana's population, whether Creole or Spanish, slave or free, prince or pirate, was, the newspaper went on to say, "with the Americans to a man."[42] Indeed, they had proven that they were all Americans.

A Nation at Last

Now there began a three-way race for Washington. In this age of plodding communications, so much had happened in December 1814 and January 1815 that would affect the whole, but the pieces lay separated by a wide ocean and scattered across half a continent—unknown to one another. From Hartford, Connecticut, Harrison Gray Otis and two fellow delegates rode south with the petitions of the Hartford Convention. While not a raised sword of secession, their message was nonetheless a call for the reprimand if not the outright dismissal of James Madison's beleaguered administration.

From New Orleans, a series of riders galloped north toward Washington with Andrew Jackson's dispatch to Secretary of War Monroe. The totality of Old Hickory's campaign at New Orleans and the subsequent British withdrawal was not yet altogether clear, but his account of that morning on the plain at Chalmette could be construed no other way than as a resounding victory. And from London, Henry Clay's secretary, Henry Carroll, and British representative Anthony St. John Baker sailed for Washington on the British sloop of war HMS *Favorite*, carrying with them for ratification by the United States Senate the official copies of the Treaty of Ghent. Which news would reach Washington first?

. . .

Meanwhile, the British government had formally ratified the Treaty of Ghent on December 28, 1814, just four days after the peace commissioners had signed it on Christmas Eve. At first blush, the treaty met with disdain in the British press, then still blissfully ignorant of the news from New Orleans. "We have retired from the combat with the stripes yet bleeding on our backs," bemoaned the *London Times*, "—with the recent defeats, at Plattsburgh, and on Lake Champlain, unavenged. To make peace at such a moment . . . betrays a deadness to the feelings of honour, and shows a timidity of disposition, inviting further insult. . . ." Indeed, the newspaper went on to say, the "inevitable consequences" of such a peace would be "the speedy growth of an American navy and the recurrence of a new and much more formidable American war. . . ." Better it was, concluded the *Times*, "that we should grapple with a young lion when he is first fleshed with the taste of our flock than await until in the maturity of his strength he bears away at once both sheep and shepherd."[1]

But clearly the British government saw it differently, as evidenced by its prompt ratification. In Vienna, where attention remained focused on Europe and the world after Napoleon, the British delegation was ecstatic. Lord Castlereagh spoke for most when he told Prime Minister Liverpool that it was a "most auspicious and seasonable event. I wish you joy of being released from the millstone of an American war."[2] Now, if only the Americans would ratify the treaty as quickly.

Vexed by storms in her attempt to reach the Chesapeake, HMS *Favorite* docked instead in New York on February 11 and found the city in a buoyant mood. Jackson's victory dispatch had reached Washington on February 4 and then been "spread north and west as fast as man and horse could carry it." Hearing the news that Carroll and Baker brought with them of a signed offer of peace, the city erupted in a bedlam of celebration. Cannon at the Battery boomed salutes; bonfires made it appear that the entire city was in flames; and torchlight parades snaked through lower Manhattan's narrow streets. By the next day, Sunday, February 12, church bells pealed the news in Philadelphia. Boston—its long ambivalence quickly forgotten—learned of the treaty on Monday and tossed aside its New England reserve to celebrate as well. No one seemed to care about the details of the peace, just that there would at long last be peace.[3]

The news from Ghent overtook the Hartford Convention delegates in Baltimore. They had already swallowed hard at the news from New Orleans, but to most this news from Ghent made them look at best like silly fools, at worst the plotting traitors some assumed they were. "The *grievance deputies* from Massachusetts and Connecticut," Winfield Scott reported to James Monroe from Baltimore, "have afforded a fine subject of jest and merriment to men of all parties."[4] Still determined to extract at least some reimbursement of New England's defense costs, the trio continued on to Washington. James Madison, with probably more than a little smirk, declined to receive them.

Henry Carroll and Anthony St. John Baker also continued on to Washington with the British ratification of the treaty, but upon their arrival in the capital city they were met with a surprise. Christopher Hughes, the secretary of the American delegation at Ghent, had sailed directly for Annapolis, weathered the storms off the Virginia capes, and arrived in Washington three days before with a copy of the treaty. Acting as expeditiously as their British counterparts, the United States Senate had ratified it on February 16 by a vote of 35–0. It was a far cry from the lengthy, contentious, and divided debate that had led to war two and a half years before. Now, in the late hours of February 17, 1815, Baker hastened to meet with President Madison and exchange the official ratifications.[5] The War of 1812 was over—or was it?

The war "has been waged," President Madison told Congress in his message the next day, "with a success which is the natural result of the wisdom of the Legislative councils, of the patriotism of the people, of the public spirit of the militia, and of the valor of the military and naval forces of the country."[6] Those who chose to pick apart the president's statement found ample fodder. Had not there been a divided Congress; a whole region ambivalent, some of its inhabitants even trading with the enemy; countless militia that repeatedly refused to engage or leave their native state; and a chain of command and control that only recently had shed itself of petty and pompous pontificators? Yes indeed, on all counts, but who wanted to remember that now? "Who would not be an American?" asked Hezekiah Niles's *Weekly Register*. No one, was the unanimous reply. "Long live the republic!"[7]

Meanwhile, back on the other side of the Atlantic, Great Britain waited eagerly for news of the American ratification and the latest word from Admiral Cochrane on the Gulf Coast. After all, the Treaty of Ghent had called for a return to *status ante bellum*. Great Britain had never recognized Napoleon's coercive dealings with Spain and his subsequent sale of Louisiana to the United States. If the Royal Navy should by now find itself ensconced in New Orleans, the British government, for all of its interest in ending the American war, might well claim this activity decidedly outside the intents of the treaty. Even British prime minister Lord Liverpool wrote to the Duke of Wellington as late as February 28, 1815, noting that "it is very desirable that the American war should terminate with a brilliant success on our part."[8] Was the war over or not?

Such musings on the fate of New Orleans and the legal status of Louisiana evaporated on the wind on March 9 when news reached London that it was General Jackson who occupied the Crescent City and not Admiral Cochrane. There was also news that Napoleon had somehow managed to escape from Elba and was once more marching on Paris. Suddenly the prompt resolution of the American war was again of great importance to the British. Four days later, news of the American ratification came to London. Lord Liverpool and his government breathed a collective sigh of relief and—New Orleans or not—turned their full attention back to the continent of Europe. The War of 1812 was indeed over.

• • •

So what had the war accomplished? In the spring of 1812, before war was declared, John C. Calhoun of South Carolina wrote to his cousin, Patrick Noble, asserting that the coming war "will be a favorite one with the country. Much honor awaits those who may distinguish themselves."[9] Calhoun was certainly wrong on the first point, but there was much truth in the second. Many of its participants would be heard from again and again. Some, history just couldn't seem to get rid of. General James Wilkinson danced with the devil and escaped censure in yet another court-martial, this one for the failed St. Lawrence campaign. Wilkinson then headed back to the intrigue of the Southwest, wrote a three-volume autobiographical apologia, and died in Mexico in 1825. Others were more deserving, if no less self-centered.

Winfield Scott received the news of peace one rank shy of his goal. Rumors abounded that Congress had planned to appoint a lieutenant general from among the three ablest and most deserving major generals: Andrew Jackson, Jacob Brown, and Scott himself. Now that was not to be, and a disappointed Scott wrote disingenuously to Monroe asking "what reward a government can bestow" on Jackson and Brown, if not the higher grade, rank being but "the first wish of a solider."[10]

Monroe must have seen through Scott's thinly veiled appeal for himself, but the twenty-nine-year-old major general had plenty of time. A quarter of a century would pass, but in 1841, Scott would become general-in-chief of the American army, succeeding Alexander Macomb, the hero of Plattsburgh, in that position. Later, a short and decisive campaign in the Mexican War earned Scott the Whig nomination for president in 1852, but he was trounced by a political unknown named Franklin Pierce. In 1861 Winfield Scott would still command the Union armies in the opening round of a far more bitter contest until pushed aside by the new order in the boots of braggart George B. McClellan. By then Scott had seen too much of war and was a far cry from the eager young officer who had momentarily seized Queenston Heights in his first engagement.

Two other generals did ride the War of 1812 to the presidency, although one arrived there so old and infirm that he survived the office but a month. Beating out Henry Clay for the Whig nomination in 1840, William Henry Harrison, the hero of that relatively minor engagement near the banks of the Tippecanoe, was given quiet John Tyler of Virginia as his running mate, and rode to victory with a slogan of "Tippecanoe and Tyler, too." A month after his inauguration, Harrison was dead of pneumonia.

The other general, of course, was Andrew Jackson. Old Hickory received a plurality of the popular vote for president three times, being denied the office the first time in 1824 after failing to win a majority of the electoral college. The victor then was John Quincy Adams, thanks in part to the support of Henry Clay, who became his secretary of state. Jackson defeated Adams's bid for reelection in 1828 and won again in 1832 against none other than Clay.

Between that day on the plain at Chalmette and the 1824 election, Jackson took the lead in wrestling the rest of Florida away from Spain. Jackson biographer Robert Remini recounts the story that upon hearing of his nomination for president, a woman in North Carolina where he had grown up exclaimed incredulously, "What! Jackson up for the President? *Jackson? Andrew Jackson?* The Jackson that used to live in Salisbury? . . . Well if Andrew Jackson can be President, anybody can!"[11] Such was the power of the enduring fame of the Battle of New Orleans.

For his part, Henry Clay was reelected to his position as Speaker of the House upon his return from Ghent. Interrupted only by his cabinet service, he remained a power in the House and Senate until his death in 1852. Clay managed to hold the Union together while he lived, but never achieved his ultimate dream of the presidency.

Of Clay's fellow commissioners at Ghent, John Quincy Adams came home to serve eight years as James Monroe's secretary of state. James A. Bayard returned home to Wilmington, Delaware, and though only forty-eight, died six days later. Albert Gallatin remained an adviser to presidents, banker and financial expert, and respected elder statesman until his death in 1849. Jonathan Russell never rose above his secondary status in the group and secured his place there by forging documents of the Ghent conference in a later political dispute with Adams.

A host of other political and military leaders made their reputation or cut their teeth by serving in the War of 1812. Richard Mentor Johnson became Martin Van Buren's vice president in 1837, even if he had not been the one to kill Tecumseh. David Farragut began his naval career as a thirteen-year-old midshipman on the USS *Essex* and would later damn the torpedoes and steam into Mobile Bay. In between them was a host of governors, senators, congressmen, and the neighbor next door who were quick to boast that, yes, of course, they had served, and who were just as quick to forget the grimmer moments of the conflict.

On the British side, the best and bravest of His Majesty's service who had come to North America lay dead: Brock, Ross, and Pakenham. The last's brother-in-law, the Duke of Wellington, was undoubtedly quite glad that he had politely but firmly refused the overtures of his government to accept a North American role. Within three months of news of the American ratification of the Treaty of Ghent, Wellington ensured himself immortality by once and for all defeating Napoleon at Waterloo. Major General John Lambert erased the stain of his role at New Orleans by hastening to Waterloo and commanding Wellington's Tenth Brigade with such aplomb that he was decorated by Parliament as well as its Russian and Austrian allies.

Lieutenant Colonel Thomas Mullens of the Forty-fourth Regiment, who had suddenly disappeared from the field on that fateful morn outside New Orleans, was summoned before a court-martial in Dublin. It found that he had "shamefully neglected and disobeyed" orders given to him by General Gibbs and dismissed him from the service for his "scandalous and infamous misbehavior." One fellow British officer commented: "It was all very well to victimize old Mullens; the fascines, ladders, etc. could have been supplied by one word which I will not name."[12] That word, of course, was "courage."

Sir George Prevost returned to Great Britain to face a court-martial for his conduct at Plattsburgh, but died before it was to convene in January 1816. Admiral Alexander Cochrane escaped such scrutiny, but after his return to England, he languished without a command for many years. Admiral George Cockburn was planning another pillage and plunder expedition—this one against Savannah, Georgia—when he received news of the peace. Cockburn's next assignment was to convey Napoleon aboard HMS *Northumberland* to his final exile at St. Helena and to serve for a time as governor of that little domain. Until his death in 1853, Cockburn held a host of military and political posts, including serving in Parliament and as First Lord of the Admiralty.

So what had the war accomplished? Upon hearing the news from Ghent, Rosalie Stier Calvert, mistress of a Maryland plantation, expressed relief and claimed that "now, if we can just get rid of our Democratic administration and have a president of the Federalist party, the United States would soon recover from the losses they have suffered."[13] But that was never going to happen.

In the dismal December of the war's second year, President James Madison had addressed Congress and tried to put the best possible light on the trials and tribulations the nation had endured on the battlefields. "The war, with its vicissitudes," Madison proclaimed, "is illustrating the capacity and the destiny of the United States to be a great, a flourishing, and a powerful nation."[14] Now, whatever Madison's shortcomings as a charismatic leader, he had been proven correct, and he could justly don the cloak of victor. There was to be no better evidence of that than James Monroe's election to the presidency in 1816 as Madison's successor. John Armstrong might grumble about "the Virginia Dynasty," but it was fact.

In the election of 1816, only the states of Massachusetts, Connecticut, and Delaware supported the Federalist candidate, Rufus King of New York. The Federalist Party was indeed dead, buried under the tombstone of the Hartford Convention. Monroe's goal, the president-elect wrote Andrew Jackson shortly after the election, was "to prevent the reorganization and revival of the federal party." Monroe thought "that the existence of parties is not necessary to free government" and in any event assumed "that the great body of the federal party are republican. . . ."[15] That proposition certainly held true in 1820, when Monroe was reelected without opposition, but it would shortly unravel in the four-way race of 1824.

In the meantime, James Monroe was able to preside over a national coalescence that had been tested and fired at places such as Put-in Bay, Lundy's Lane, and Plattsburgh. In the summer of 1823, scarcely eleven years after he had waited expectantly for British peace commissioners to appear in Ghent, John Quincy Adams, now Monroe's secretary of state, spoke for a nation much surer of itself than it had been then. In the face of Russian territorial claims that were extending down the Pacific coast as far as Fort Ross, California, Adams firmly informed the Russian ambassador in Washington that the United States would strongly contest the right of Russia to *any* new territorial claims in North America.

Six months later, Adams's words were broadened and incorporated into President Monroe's annual message to Congress. "The American continents," Monroe asserted, "by the free and independent condition which they have assumed and maintain, are henceforth not to be considered as subject for future colonization by any European power." Monroe went on to warn European governments to keep their hands off the affairs of the newly independent governments of their former colonies in Latin America. These noncolonization and nonintervention statements became the twin cornerstones of what came to be called the Monroe Doctrine. James Monroe, too, had come of age with the nation.[16]

• • •

So what had the war accomplished? One American historian later wrote that "the bonfires, the cannon, the church bells which celebrated the Peace of Ghent constituted less a shout of triumph than a sigh of relief."[17] At first, that was definitely true. There was an economy to rebuild, and peace quickly proved a boon to the commerce of both the United States and Great Britain. But once this economic resurgence was in progress, the sigh of relief rapidly turned to a swagger of triumph. Americans quickly forgot the war's disheartening defeats and chose instead to remember a few proud moments: Lawrence's dying words of "Don't give up the ship," the glories of "Old Ironsides," and of course, Jackson's stand before New Orleans.

Great Britain brushed aside its disaster at New Orleans and after Waterloo was content to glory in its Napoleonic laurels. But for the United States, the triumph of Andrew Jackson's roughshod collection of army regulars, backwoods militia, and bayou pirates over the elite of the British Empire came to fill a huge void in the American psyche—not only propelling Jackson to the presidency, but affirming America on the course that would extend its borders to the Pacific.

Unwittingly or not, Jackson himself touched on this feeling in a proclamation he ordered read at the head of each corps of his army near New Orleans on January 21, 1815. Praising the "undaunted courage, patriotism, and patience, under hardships and fatigues," of his troops, Jackson proclaimed: "Natives of different states, acting together, for the first time, in this camp; differing in habits and in language, instead of viewing in these circumstances the germ of distrust and division, you have made them the source of an honourable emulation, and from the seeds of discord itself have reaped the fruits of an honourable union."[18]

In those words, Jackson grasped at what his biographer, Robert Remini, said the War of 1812 had been about. "In a real sense," wrote Remini, "the War of 1812 was part of a search for national identity."[19] And in its aftermath, Jackson and his contemporaries seemed to sense that it had been found as a nation.

"The United States *have* . . .," James Madison had written to Congress upon the brink of war, employing the plural verb for the nation. "The United States and *their* territories . . . ," intoned Congress upon declaring it, again employing the plural.[20] Somehow, in the caldron of war with—to paraphrase Madison—all of "its vicissitudes," the *United States* as a plural term for eighteen disparate states had become a singular term for one united nation. To be sure, there would still be strong regional loyalties and nagging issues of states' rights, but at home and abroad after 1815, there was a newfound pride of *national* identity.

"In every aspect I must acknowledge that the war has been useful," wrote Albert Gallatin to Thomas Jefferson late in 1815. "The character of America stands now as high as ever on the European continent, and higher than ever it did in Great Britain."[21]

A few months later, Gallatin was even more specific: "The war has renewed and reinstated the national feelings and character which the Revolution had given, and which were daily lessened. The people now have more general objects of attachment with which their pride and political opinions are connected. They are more Americans; they feel and act more as a nation; and I hope that the permanency of the Union is thereby secured."[22]

So what had the war accomplished? With plenty of missteps, the United States had cast aside its cloak of colonial adolescence and stumbled forth onto the world stage. To be sure, there would be family quarrels—one of which would threaten to tear it asunder two generations hence. But after the War of 1812, the United States was a singular term, not plural. After the War of 1812, there was no longer any doubt that the United States of America would become a force to be reckoned with in North America and in time throughout the world. The war had forged a nation.

Endnotes

BOOK ONE: Drumbeats (1807–1812)

To Steal an Empire

1. Milton Lomask, *Aaron Burr: The Conspiracy and Years of Exile, 1805-1836* (New York: Farrar, Straus, Giroux, 1982), p. 58.
2. Thomas Fleming, *Duel: Alexander Hamilton, Aaron Burr and the Future of America* (New York: Basic Books, 1999), pp. 93–94.
3. Harold C. Syrett, ed., *The Papers of Alexander Hamilton*, vol. 25, July 1800–April 1802 (New York: Columbia University Press, 1977), p. 257 (Hamilton to Wolcott, December 16, 1800).
4. Fleming, *Duel*, pp. 94, 100, 102.
5. Ibid., p. 189.
6. Ibid., pp. 63-65.
7. Ibid., pp. 146-47.
8. Ibid., pp. 67, 139, 143.
9. John M. Taylor, "An Accomplished Villain," *American History Illustrated* vol. 13, no. 9 (January 1979): 4–9; Fleming, *Duel*, p. 65, 140.
10. Harold C. Syrett, ed., *The Papers of Alexander Hamilton*, vol. 26, May 1, 1802–October 23, 1804 (New York: Columbia University Press, 1979), p. 173 (Wilkinson to Hamilton, November 15, 1803).
11. Harold C. Syrett, ed., *The Papers of Alexander Hamilton*, vol. 23, April 1799–October 1799 (New York: Columbia University Press, 1976), pp. 227–28 (Hamilton to McHenry, June 27, 1799).

12. Fleming, *Duel*, pp. 245–47.

13. Ibid., pp. 260–62.

14. Syrett, *Papers of Alexander Hamilton*, vol. 26, pp. 217–18. (James Wilkinson to Hamilton, March 26, 1804).

15. Mary-Jo Kline, ed., *Political Correspondence and Public Papers of Aaron Burr*, vol. 2 (Princeton, N.J.: Princeton University Press, 1983), pp. 891–92 (Merry to Harrowby, August 6, 1804).

16. Raymond E. Fitch, ed., *Breaking with Burr: Harman Blennerhassett's Journal, 1807* (Athens: Ohio University Press, 1988), p. 125.

17. Minnie Kendall Lowther, *Blennerhassett Island in Romance and Tragedy* (Rutland, Vt.: Tuttle Publishing, 1939), pp. 7, 9, 14–15.

18. Ibid., pp. x, 10–12.

19. Kline, *Political Correspondence and Public Papers of Aaron Burr*, p. 950.

20. Robert V. Remini, *Andrew Jackson and the Course of American Empire, 1767–1821* (New York: Harper & Row, 1977), pp. 145–51.

21. Ibid., pp. 152–53.

22. James Wilkinson, *Memoirs of My Own Times*. (Philadelphia: Abraham Small, 1816), vol. 2, appendix 99 (Wilkinson to Cushing, November 7, 1806).

23. Dumas Malone, *Jefferson the President: Second Term, 1805–1809* (Boston: Little, Brown, 1974), pp. 338–41.

24. Fitch, *Breaking with Burr*, p. 81; Lowther, *Blennerhasett Island*, p. 34.

25. Remini, *Jackson and the Course*, p. 157.

First Blood at Sea

1. John A. Garraty, *The American Nation: A History of the United States* (New York: Harper & Row, 1966), pp. 189–90.

2. Malone, *Jefferson the President*, pp. 416–19.

3. Paul Barron Watson, *The Tragic Career of Commodore James Barron* (New York: Coward-McCann, 1942), pp. 25, 31.

4. Malone, *Jefferson the President*, pp. 419–21.

5. Watson, *Tragic Career*, p. 35.

6. Theodore Roosevelt, *The Naval War of 1812* (Annapolis: Naval Institute Press, 1987), p. 35.

7. Malone, *Jefferson the President*, pp. 421–22.

8. Ibid., pp. 425–26.

9. Pierre Berton, *The Invasion of Canada, 1812–1813* (Toronto: McClelland and Stewart, 1980), p. 37.

10. J. Mackay Hitsman, *The Incredible War of 1812: A Military History* (Toronto: Robin Brass Studio, 1999), p. 16.
11. Roosevelt, *Naval War*, p. 35; Hitsman, *Incredible War*, p. 23.

War Hawks and Tippecanoe

1. *Niles' Weekly Register*, 2, no. 1 (March 7, 1812), p. 5.
2. Donald R. Hickey, *The War of 1812: A Forgotten Conflict* (Urbana: University of Illinois Press, 1989), p. 30; Bradford Perkins, *Prologue to War: England and the United States, 1805–1812* (Berkeley: University of California Press, 1963), pp. 261–62.
3. *Annals of the Congress of the United States*, Twelfth Congress, First Session, p. 533 (December 16, 1811).
4. R. B McAfee, "William H. Harrison," *The Library of Historic Characters and Famous Events* (Boston: J. B. Millet, 1907), vol. II, pp. 309–13; see also Freeman Cleaves, *Old Tippecanoe: William Henry Harrison and His Time* (New York: Charles Scribner's Sons, 1939).
5. Garraty, *American Nation*, p. 193; Cleaves, *Old Tippecanoe*, p. 67.
6. Logan Esarey, ed., *Messages and Letters of William Henry Harrison*, vol. 1, 1800–1811 (Indianapolis: Indiana Historical Commission, 1923), p. 389 (Harrison to Eustis, November 3, 1809); the terms of the treaty are at pp. 359–61. The $2 an acre information is from Glenn Tucker, *Poltroons and Patriots: A Popular Account of the War of 1812* (Indianapolis: Bobbs-Merrill, 1954), p. 109.
7. Cleaves, *Old Tippecanoe*, p. 11.
8. Tucker, *Poltroons and Patriots*, pp. 102–4.
9. Ibid., pp. 109–12; Cleaves, *Old Tippecanoe* pp. 72–82.
10. Esarey, *Messages and Letters of William Henry Harrison*, p. 549 (Harrison to Eustis, August 7, 1811).
11. John K. Mahon, *The War of 1812* (Gainesville: University of Florida Press, 1972), pp. 20–27; Robert Breckenridge McAfee, *History of the Late War in the Western Country*. (Ann Arbor, Mich.: University Microfilms, 1966), pp. 27–38; Cleaves, *Old Tippecanoe*, pp. 93–109; for the story of Harrison's horse see Esarey, *Messages and Letters of William Henry Harrison*, vol. 1, pp. 691–92 (Harrison to Scott, December 1811).
12. McAfee. "William H. Harrison ," p. 317.
13. Perkins, *Prologue to War*, pp. 287–88.

Mr. Madison's War

1. Hickey, *War of 1812*, pp. 17–19; Garraty, *American Nation* p. 187.
2. Curtis P. Nettles, *The Emergence of a National Economy, 1775–1815*, vol. 3 of *The Economic History of the United States* (New York: Holt, Rinehart and Winston, 1962), p. 328; Malone, *Jefferson the President*, pp. 482, 628, 651–55; Garraty, *American Nation*, pp. 191–92.
3. Garraty, *American Nation*, p. 193.
4. Hickey, *War of 1812*, p. 38; *Annals of the Congress of the United States*, Twelfth Congress, First Session, p. 1162. (March 9, 1812).
5. Hickey, *War of 1812*, p. 39; Perkins, *Prologue to War*, pp. 369–72.

Concessions Too Late

1. Hitsman, *Incredible War*, p. 45.
2. Dumas Malone, *Jefferson and His Time: The Sage of Monticello* (Boston: Little, Brown, 1977), pp. 85, 131.
3. *London Times*, December 16, 1811.
4. Henry Adams, *History of the United States of America During the Administrations of James Madison* (New York: Library of America, 1986), p. 439.
5. Perkins, *Prologue to War*, pp. 393, 395.
6. Julius W. Pratt, *A History of United States Foreign Policy* (Englewood Cliffs, N.J.: Prentice-Hall, 1955), p. 132.
7. Perkins, *Prologue to War*, pp. 393, 397, 403.
8. Gaillard Hunt, ed. *The Writings of James Madison*, vol. 8, 1808–1819 (New York: G. P. Putnam's Sons, 1908), pp. 191–201.
9. Perkins, *Prologue to War*, pp. 406–9.
10. Hickey, *War of 1812*, p. 46.
11. Perkins, *Prologue to War*, pp. 413–14.
12. Hickey, *War of 1812*, p. 75.
13. Perkins, *Prologue to War*, pp. 414–15; Mahon, *War of 1812*, p. 31.
14. Pratt, *History of United States Foreign Policy*, p. 132.
15. Perkins, *Prologue to War*, p. 415.
16. Ibid.
17. Stanislaus Murray Hamilton, ed., *The Writings of James Monroe*, vol. 5, 1807–1816 (New York: G. P. Putnam's Sons, 1901), p. 211 (Monroe to Taylor, June 13, 1812).
18. J. C. A. Stagg, *The Papers of James Madison*, Presidential Series, vol.

4, 5, November 1811–9 July 1812 (Charlottesville: University Press of Virginia, 1999), pp. 397–98 (Gerry to Madison, May 19, 1812).

BOOK TWO: Bugles (1812–1814)

Oh, Canada

1. *Annals of the Congress of the United States*, Eleventh Congress, Second Session, p. 580 (February 22, 1810); Garraty, *American Nation*, p. 197.

2. Harry L. Coles, *The War of 1812* (Chicago: University of Chicago Press, 1965), pp. 38–39.

3. Hugh Hastings, ed., *Public Papers of Daniel D. Tompkins, Governor of New York, 1807–1817*, Military–vol. 3 (Albany: J. B. Lyon, 1902), p. 27 (Tompkins to Macomb, July 12, 1812).

4. *Connecticut Courant* (Hartford), June 30, 1812.

5. Hamilton, *Writings of James Monroe*, vol. 5, p. 213.

6. Hickey, *War of 1812*, pp. 75, 78–79.

7. Ibid., p. 80.

8. McAfee, *History of the Late War*, p. 52.

9. William Hull, *Memoirs of the Campaign of the North Western Army of the United States, A.D. 1812* (Boston: True & Greene, 1824), p. 17.

10. Mahon, *War of 1812*, pp. 43–44; McAfee, *History of the Late War*, pp. 50–51; Hickey, *War of 1812*, pp. 80–81.

11. McAfee, *History of the Late War*, p. 56.

12. Hickey, *War of 1812*, p. 81.

13. McAfee, *History of the Late War*, pp. 58–60.

14. Tucker, *Poltroons and Patriots*, p. 155.

15. McAfee, *History of the Late War*, pp. 61–62.

16. Hickey, *War of 1812*, pp. 81–82.

17. Mahon, *War of 1812*, pp. 46–47.

18. David Lavender, *Fist in the Wilderness* (Garden City, N.Y.: Doubleday, 1964), pp. 5–6, 8–9.

19. Ibid., pp. 18–19, 41.

20. Hitsman, *Incredible War*, pp. 72–73.

21. Lavender, *Fist in the Wilderness*, pp. 185–87; McAfee, *History of the Late War*, pp. 70–73.

22. McAfee, *History of the Late War*, p. 70.

23. Berton, *Invasion of Canada*, pp. 191–97; Hitsman, *Incredible War*, p. 82; McAfee, *History of the Late War*, pp. 98–99.

24. McAfee, *History of the Late War*, p. 83.

25. Tucker, *Poltroons and Patriots*, p. 149.

26. Hitsman, *Incredible War*, p. 80.

27. Ibid., pp. 81, 327; McAfee, *History of the Late War*, p. 89.

28. Hickey, *War of 1812*, p. 84; McAfee, *History of the Late War*, p. 89.

29. Hickey, *War of 1812*, p. 84; McAfee, *History of the Late War*, pp. 89, 92, 97; "unsoldierly alacrity" quote in "William Hull," *Funk & Wagnalls New Standard Encyclopedia*, (New York: Unicorn Press, 1947), vol. 16, p. 37.

30. Hickey, *War of 1812*, p. 85.

31. Ibid., p. 86.

32. Ibid.

33. Hitsman, *Incredible War*, pp. 94–96.

34. John S. D. Eisenhower, *Agent of Destiny: The Life and Times of General Winfield Scott* (New York: Free Press, 1997), pp. 1, 8, 10, 18.

35. Ibid., pp. 27, 29, 36–39.

36. Hickey, *War of 1812*, p. 87, Eisenhower, *Agent of Destiny*, p. 39.

37. Hitsman, *Incredible War*, pp. 98–100; Eisenhower, *Agent of Destiny*, p. 40.

38. Hickey, *War of 1812*, pp. 87–88.

39. Ibid., p. 88; Tucker, *Poltroons and Politics*, p. 146.

40. James F. Hopkins, ed. *The Papers of Henry Clay*, vol. 1., The Rising Statesman, 1797–1814 (Lexington: University of Kentucky Press, 1959), p. 824 (Clay to Bodley, December 18, 1812).

41. Hickey, *War of 1812*, p. 90.

Hurrah for Old Ironsides

1. Bruce Grant, *Isaac Hull: Captain of Old Ironsides* (Chicago: Pellegrini and Cudahy, 1947), pp. 7, 11–15, 26.

2. Roosevelt, *Naval War*, p. 67; Harry Hansen, *Old Ironsides: The Fighting Constitution* (New York: Random House, 1955), p. 13.

3. Grant, *Isaac Hull*, pp. 26, 57, 158, 163, 165.

4. See Roosevelt, *Naval War*, pp. 69, 72–86 for a complete discussion of classifications, guns, and the weight of a broadside.

5. Hickey, *War of 1812*, p. 92.

6. Ibid., pp. 92–93.

7. Grant, *Isaac Hull*, pp. 194–95, 202; 225; Roosevelt, *Naval War*, p. 95.

8. Grant, *Isaac Hull*, pp. 207–16; Roosevelt, *Naval War*, pp. 95–99.

9. Grant, *Isaac Hull*, p. 391.

10. Ibid., pp. 233–237; Roosevelt *Naval War*, 101–07.

11. Grant, *Isaac Hull*, p. 239.

12. Ibid., pp. 240–42; Roosevelt, *Naval War*, p. 109.

13. *Columbian Centinel* (Boston), September 2, 1812.

14. Grant, *Isaac Hull*, p. 253.

15. *London Times*, October 9, 1812.

16. A. T. Mahan, *Sea Power in Its Relations to the War of 1812* (Boston: Little, Brown, 1905), vol. I, pp. 334–35.

17. Grant, *Isaac Hull*, p. 4.

18. *London Times*, October 29, 1812.

19. Roosevelt *Naval War*, 117–127; C. S. Forester, *The Age of Fighting Sail: The Story of the Naval War of 1812* (Garden City, N.Y.: Doubleday, 1956), pp. 107–11; "grapeshot and canister" quote in Edgar Stanton Maclay, *A History of the United States Navy*, vol. 1, (New York: D. Appleton, 1910), pp. 378–79.

20. Roosevelt, *Naval War*, p. 127.

21. Forester, *Age of Fighting Sail*, p. 112; Grant, *Isaac Hull*, p. 260.

22. Roosevelt, *Naval War*, pp. 128–38; Forester, *Age of Fighting Sail*, pp. 112, 118–22.

23. Forester, *Age of Fighting Sail*, p. 127.

Marching on a Capital

1. Mahan, *Sea Power*, vol. 1, p. 361.

2. Robert Malcomson, *Lords of the Lake: The Naval War on Lake Ontario, 1812–1814* (Annapolis: Naval Institute Press, 1998), pp. 38–42.

3. William S. Dudley, ed., *The Naval War of 1812: A Documentary History*, vol. 1, 1812 (Washington, D.C.: Naval Historical Center, 1985), pp. 307–8 (Hamilton to Chauncey, September 11, 1812).

4. Malcomson, *Lords of the Lake*, p. 328.

5. Dudley, *Naval War*, vol. 1, p. 316 (Chauncey's Report to Hamilton, September 26, 1812).

6. David Curtis Skaggs and Gerard T. Altoff, *A Signal Victory: The Lake Erie Campaign, 1812–1813* (Annapolis: Naval Institute Press, 1997), p. 41.

7. Mahan, *Sea Power* vol. 1, pp. 354–55.

8. Dudley, *Naval War*, vol. 1, p. 322 (Dobbins to Elliott, October 11, 1812).

9. Malcomson, *Lords of the Lake*, p. 81.

10. Hickey, *War of 1812*, p. 127; Malcomson, *Lords of the Lake*, p. 84.

11. Pierre Berton, *Flames Across the Border, 1813–1814* (Toronto: McClelland and Stewart, 1981), pp. 29–30; Malcomson, *Lords of the Lake*, pp. 87–89; Skaggs and Altoff, *Signal Victory*, pp. 38–39.

12. Benson J. Lossing, *The Pictorial Field-Book of the War of 1812* (New York: Harper & Brothers, 1869), p. 586.

13. Ibid., p. 587.

14. Tucker, *Poltroons and Patriots*, p. 245.

15. Berton, *Flames*, pp. 45–52; Malcomson, *Lords of the Lake*, pp. 103–12; Tucker, *Poltroons and Patriots*, pp. 242–56; Hitsman, *Incredible War*, pp. 136–41; Hickey, *War of 1812*, p. 129.

16. Malcomson, *Lords of the Lake*, p. 109.

17. E. Cruikshank, *The Documentary History of the Campaign upon the Niagara Frontier in the Year 1813* (Welland, Ontario: Lundy's Lane Historical Society, 1902), part 5, pp. 281–82.

18. Adams, *History of United States*, p. 735 (Brown to Dearborn, July 25, 1813).

19. Eisenhower, *Agent of Destiny*, p. 51.

20. Ibid., pp. 56–59; Hickey, *War of 1812*, p. 139.

21. Mahon, *War of 1812*, pp. 149–50.

22. Hitsman, *Incredible War*, p. 155.

23. Berton, *Flames*, pp. 82–84, 86, 91; Hitsman, *Incredible War*, pp. 154–55.

24. Hickey, p. 141.

Don't Give Up the Ship

1. Roosevelt, *Naval War*, pp. 166–70.

2. Tucker, *Poltroons and Patriots*, pp. 259–61.

3. Mahan, *Sea Power*, vol. 2, pp. 131–32.

4. Roosevelt, *Naval War*, p. 178.

5. Ibid., pp. 177–79.

6. Ibid., p. 179.

7. Ibid., p. 179; Tucker, *Poltroons and Patriots*, pp. 264–65

8. Roosevelt, *Naval War*, pp. 180–82; Mahan, *Sea Power*, vol. 2, pp. 137–39.

9. Tucker, *Poltroons and Patriots*, p. 267.

10. Mahan, *Sea Power*, vol. 2, p. 138; Roosevelt, *Naval War*, p. 182.

11. Mahan, *Sea Power*, vol. 2, p. 139.

12. Ibid., pp. 139–40.

13. *London Morning Chronicle*, July 9, 1813.

14. Roosevelt, *Naval War*, pp. 185–86.

We Have Met the Enemy

1. Richard Dillon, *We Have Met the Enemy: Oliver Hazard Perry: Wilderness Commodore* (New York: McGraw Hill, 1978), pp. 98–99.

2. Skaggs and Altoff, *Signal Victory*, pp. 45–46.

3. Tucker, *Poltroons and Patriots*, pp. 306–11.

4. Dudley, *Naval War*, vol. 1, p. 332 (Brock to Prevost, October 22, 1812).

5. Skaggs and Altoff, *Signal Victory*, pp. 52–53.

6. Dudley, *Naval War*, vol. 2, p. 545 (Prevost to Barclay, July 21, 1813).

7. Skaggs and Altoff, *Signal Victory*, pp. 61–62, 187; Berton, *Flames*, p. 155.

8. Tucker, *Poltroons and Patriots*, p. 306.

9. Mahan, *Sea Power*, vol. 2, pp. 62–63.

10. Lossing, *Pictorial Field-Book*, p. 515.

11. Dudley, *Naval War*, vol. 2, p. 546 (Perry to Jones, August 4, 1813); Mahan, *Sea Power*, vol. 2, pp. 69–73; Skaggs and Altoff, *Signal Victory*, pp. 84–86.

12. Dudley, *Naval War*, vol. 2, p. 547 (Barclay to Yeo, August 5, 1813); Skaggs and Altoff, *Signal Victory*, pp. 66, 68.

13. Mahan, *Sea Power*, vol. 2, p. 77.

14. Ibid., p. 75.

15. Ibid., p. 76.

16. Ibid., p. 77; Roosevelt, *Naval War*, pp. 242–43.

17. Mahan, *Sea Power*, vol. 2, pp. 80–81.

18. Ibid., p. 82. For a detailed discussion of Perry's orders and Elliott's sailing, see pp. 83–88.

19. Dudley, *Naval War*, vol. 2, p. 556 (Barclay to Yeo, September 12, 1813).

20. Mahan, *Sea Power*, vol. 2, p. 93; Skaggs and Altoff, *Signal Victory*, pp. 141–45.

21. Alfred Brunson, *A Western Pioneer; or, Incidents in the Life and Times of Alfred Brunson* (Cincinnati: Hitchcock, Walden, 1872), vol. 1, p. 130; Mahan, *Sea Power*, vol. 2, p. 94.

22. Dudley, *Naval War*, vol. 2, pp. 553–54 (Perry to Harrison, Perry to Jones, September 10, 1813).

23. Roosevelt, *Naval War*, p. 254.

24. Dudley, *Naval War*, vol. 2, p. 558 (Perry to Jones, September 13, 1813); Skaggs, and Altoff, *Signal Victory*, pp. 148, 155, 158.

25. Skaggs and Altoff, *Signal Victory*, pp. 164, 167.

26. Ibid., p. 180.

27. Ibid., p. 181.

28. Dillon, *We Have Met the Enemy*, pp. 212, 214–15; Tucker, *Poltroons and Patriots*, p. 368, Skaggs and Altoff, *Signal Victory*, p. 182.

Old Hickory Heads South

1. Remini, *Jackson and the Course*, pp. 168–69.

2. George Dangerfield, *The Era of Good Feelings* (New York: Harcourt, Brace, 1952), p. 127.

3. Remini, *Jackson and the Course*, p. 170.

4. James Parton, *Life of Andrew Jackson* (New York: Mason Brothers, 1860), vol. 1, p. 372.

5. Remini, *Jackson and the Course*, pp. 172–73.

6. James R. Jacobs, *Tarnished Warrior: The Story of Major-General James Wilkinson*. (New York: MacMillan, 1938), p. 279; Parton, *Life of Andrew Jackson*, vol. 1, pp. 377–81.

7. McAfee, *History of the Late War*, p. 457.

8. Jacobs, *Tarnished Warrior*, p. 281.

9. Remini, *Jackson and the Course*, p. 180.

10. Ibid., p. 182.

11. Ibid., p. 184.

12. Ibid., pp. 184–86.

13. R. David Edmunds, *Tecumseh and the Quest for Indian Leadership* (New York: Little, Brown, 1984), p. 150.

14. Ibid., pp. 146–53.

15. McAfee, *History of the Late War*, p. 458.

16. Mahon, *War of 1812*, pp. 231–32.

17. McAfee, *History of the Late War*, pp. 461–63; Mahon, *War of 1812*, pp. 234–35; Tucker, *Poltroons and Patriots*, pp. 448–51.

18. *Niles's Weekly Register* 5, no. 7, (October 16, 1813), p. 105.

19. Remini, *Jackson and the Course*, pp. 191–92.

20. McAfee, *History of the Late War*, p. 465; Mahon, *War of 1812*, pp. 236–37.

21. Remini, *Jackson and the Course*, pp. 196–97; Mahon, *War of 1812*, p. 237.

22. Mahon, *War of 1812*, p. 240; Remini, *Jackson and the Course*, p. 206; McAfee, *History of the Late War*, p. 472.

23. Remini, *Jackson and the Course*, pp. 208–10.

24. Ibid., pp. 213–17.

25. Hickey, *War of 1812*, p. 151.

26. Remini, *Jackson and the Course*, p. 218.

27. Ibid., p. 232.

28. Bureau of American Ethnology, *Annual Report*, 1897–98 (Washington, D.C.: Government Printing Office, 1900), vol. 1, p. 97.

On the Thames and St. Lawrence

1. Hickey, *War of 1812*, p. 135.

2. Mahon, *War of 1812*, p. 160.

3. Ibid., pp. 159–60; Cleaves, *Old Tippecanoe*, pp. 164–66.

4. McAfee, pp. *History of the Late War*, 322–29; Mahon, *War of 1812*, pp. 163–65; Cleaves, *Old Tippecanoe*, pp. 180–82.

5. Hickey, *War of 1812*, p. 136.

6. McAfee, *History of the Late War*, p. 337; Mahon, *War of 1812*, p. 178.

7. Coles, *War of 1812*, p. 129.

8. Hickey, *War of 1812*, p. 137; Mahon, *War of 1812*, pp. 181–85.

9. Esarey, *Messages and Letters of William Henry Harrison*, vol. 1, p. 549.

10. Mahon, *War of 1812*, p. 184.

11. Edmunds, *Tecumseh*, p. 216.

12. Jacobs, *Tarnished Warrior*, pp. 284–86.

13. Mahon, *War of 1812*, p. 205.

14. Jacobs, *Tarnished Warrior*, pp. 288–89; Eisenhower, *Agent of Destiny*, p. 67.

15. Wilkinson, *Memoirs*, vol. 3, appendix 1, p. 2.

16. Hickey, *War of 1812*, pp. 143–45.

17. Hitsman, *Incredible War*, pp. 184–87.

18. Hickey, *War of 1812*, p. 145.

19. Jacobs, *Tarnished Warrior*, p. 289.

20. Ibid., p. 293.
21. Donald E. Graves, *Field of Glory: The Battle of Crysler's Farm, 1813* (Toronto: Robin Brass Studio, 1999), p. 223.
22. Ibid., pp. 224–25.
23. Jacobs, *Tarnished Warrior*, p. 296; Hitsman, *Incredible War*, p. 191.
24. Graves, *Field of Glory*, p. 272.
25. Ibid., p. 295.
26. Tucker, *Poltroons and Patriots*, p. 418 (*New York Gazette*, January 9, 1814).
27. Berton, *Flames*, pp. 255–57.
28. Hickey, *War of 1812*, pp. 142–43
29. Tucker, *Poltroons and Patriots*, p. 427 (*New York Gazette*, January 9, 1814).

The Lion's Roar

1. Hickey, *War of 1812*, pp. 152–53.
2. Dudley, *Naval War*, vol. 2, pp. 14–15 (Croker to Warren, January 9, 1813).
3. Mahan, *Sea Power*, vol. 1, p. 287.
4. Hickey, *War of 1812*, pp. 152–53.
5. Roosevelt, *Naval War*, pp. 262, 293; Dudley, *Naval War*, vol. 2, pp. 48–49, 300; Mahon, *War of 1812*, p. 110.
6. Dudley, *Naval War*, vol. 1, p. 528 (Porter to Hamilton, October 14, 1812).
7. Mahan, *Sea Power*, vol. 2, pp. 245–52; Mahon, *War of 1812*, p. 250; Roosevelt, *Naval War*, pp. 268–81.
8. *Caledonian Mercury* (Edinburgh), August 3, 1812; *London Times*, July 31, 1812.
9. *London Times*, May 24, 1814.
10. Mahan, *Sea Power*, vol. 2, pp. 330–31.
11. Hickey, *War of 1812*, p. 182.

BOOK THREE: Finale (1814–1815)

Niagara's Thunder

1. Hickey, *War of 1812*, pp. 183–85.
2. Eisenhower, *Agent of Destiny*, pp. 73–78.

3. Ibid., pp. 80–84.

4. Winfield Scott, *Memoirs of Lieut.-General Scott, LL.D.* (Freeport, N.Y.: Books for Libraries Press, 1970), vol. 1, p. 134.

5. Roosevelt, *Naval War*, p. 328.

6. Ibid., p. 328.

7. Donald E. Graves, *Where Right and Glory Lead! The Battle of Lundy's Lane, 1814* (Toronto: Robin Brass Studio, 1997), pp. 51–56, 103; Eisenhower, *Agent of Destiny*, pp. 86–89.

8. Graves, *Where Right and Glory Lead*, pp. 114–16; Eisenhower, *Agent of Destiny*, pp. 89–90.

9. Scott, *Memoirs*, vol. 1, pp. 139, 141.

10. Graves, *Where Right and Glory Lead*, pp. 139–41; Berton, *Flames*, p. 334.

11. Hickey, *War of 1812*, p. 188; Graves, *Where Right and Glory Lead*, pp. 149–53.

12. Cruikshank, *Documentary History*, vol. 1, pp. 105–06; Hickey, *War of 1812*, p. 188: Graves, *Where Right and Glory Lead*, p. 154.

13. Graves, *Where Right and Glory Lead*, p. 160.

14. Ibid., pp. 179–84, 187–88; Eisenhower, *Agent of Destiny*, pp. 93–95.

15. Graves, *Where Right and Glory Lead*, pp. 195–97; Mahan, *Sea Power*, vol. 2, p. 313.

16. Graves, *Where Right and Glory Lead*, pp. 211–12.

17. *Ibid.*, pp. 214, 219–20.

18. Mahan, *Sea Power*, vol. 2, p. 312.

Lake Champlain

1. Hitsman, *Incredible War*, pp. 289–90 (full text of letter of Earl Bathurst to Sir George Prevost, June 3, 1814).

2. Mahan, *Sea Power*, vol. 2, pp. 362–63.

3. Hickey, *War of 1812*, p. 190.

4. Mahan, *Sea Power*, vol. 2. pp. 363–64.

5. Hitsman, *Incredible War*, pp. 253–54.

6. Mahan, *Sea Power*, vol. 2, pp. 366–67.

7. "Thomas Macdonough," *Dictionary of American Biography* (New York: Charles Scribner's Sons, 1943), vol. 12, pp. 19–21; David G. Fitz-Enz, *The Final Invasion: Plattsburgh, the War of 1812's Most Decisive Battle* (New York: Cooper Square Press, 2001), pp. 39, 42.

8. Roosevelt, *Naval War*, pp. 338–40; Mahan, *Sea Power*, vol. 2, pp. 360–61.

9. Malcomson, *Lords of the Lake*, p. 242; Fitz-Enz, *Final Invasion*, pp. 58, 94–95.

10. Roosevelt, *Naval War*, p. 340–42; Mahan, *Sea Power*, vol. 2, pp. 362, 372.

11. Roosevelt, *Naval War*, pp. 347–48.

12. Hitsman, *Incredible War*, p. 255.

13. Ibid., pp. 256–257; Mahan, *Sea Power*, vol. 2, p. 374.

14. Hitsman, *Incredible War*, p. 257.

15. Mahan, *Sea Power*, vol. 2, pp. 373–75.

16. Rodney Macdonough, *The Life of Commodore Thomas Macdonough, U.S. Navy* (Boston: Fort Hill Press, 1909), p. 176.

17. Mahan, *Sea Power*, vol. 2, pp. 376–81; Roosevelt, *Naval War*, pp. 348–55; Mahon, *War of 1812*, pp. 324–25; Fitz-Enz, *Final Invasion*, p. 213; Hickey, *War of 1812*, p. 193.

18. Hitsman, *Incredible War*, p. 260.

19. Ibid., p. 262.

20. Mahan, *Sea Power*, vol. 2, p. 370; Hitsman, *Incredible War*, p. 267.

21. Hickey, *War of 1812*, p. 193; Roosevelt, *Naval War*, pp. 356–57.

22. Roosevelt, *Naval War*, p. 254.

23. Charles G. Muller, *The Proudest Day: Macdonough on Lake Champlain*, (New York: John Day, 1960), p. vii.

Another Capital Burns

1. Hickey, *War of 1812*, pp. 174, 214–15.

2. "Sir Alexander Forester Inglis Cochrane," *Dictionary of National Biography* (London: Macmillan, 1908), vol. 4, pp. 615–16; Mahan, *Sea Power*, vol. 2, p. 330.

3. *Niles' Weekly Register* 6, no. 19 (July 9, 1814), p. 317, Hickey, *War of 1812*, p. 215.

4. Adams, *History of the United States*, pp. 998–99.

5. "Sir George Cockburn," *Dictionary of National Biography*, vol. 14, pp. 640–41.

6. *London Times*, October 29, 1812.

7. "Robert Ross," *Dictionary of National Biography*, vol. 17, pp. 274–75.

8. Adams, *History of the United States*, pp. 997–98.

9. *Niles' Weekly Register* 4, no. 25 (August 21, 1813), p. 402.

10. Mahon, *War of 1812*, p. 291.

11. Mahan, *Sea Power*, vol. 2, pp. 340–41.

12. *New York Post*, August 8, 1814.
13. Mahon, *War of 1812*, p. 291–92; Jeanne T. Heidler and David S. Heidler, eds., *Encyclopedia of the War of 1812* (Santa Barbara, Calif: ABC-Clio, 1997), pp. 442–43, 558–59.
14. *Federal Republican* (Georgetown), August 12, 1814.
15. Adams, *History of the United States*, p. 993.
16. Tucker, *Poltroons and Patriots*, p. 520.
17. G. R. Gleig, *The Campaigns of the British Army at Washington and New Orleans* (Totowa, N.J.: Rowman and Littlefield, 1972), p. 54.
18. Mahon, *War of 1812*, pp. 294–95, 298–99; Adams, *History of the United States*, pp. 1006–07; 1010.
19. Adams, *History of the United States*, p. 1016.
20. Ibid., p. 1018.
21. Charles Ball, *A Narrative of the Life and Adventures of Charles Ball, A Black Man* (New York: John S. Taylor, 1837), p. 468.
22. Catherine Austin, ed., "Letters of Elbridge Gerry, Jr.," *Massachusetts Historical Society Proceedings* 47 (June 1914), p. 511; Mahon, *War of 1812*, p. 300; Adams, *History of the United States*, pp. 1011–12.
23. Tucker, *Poltroons and Patriots*, p. 537.
24. Ibid., p. 550; for casualty figures see Anthony S. Pitch, *The Burning of Washington: The British Invasion of 1814* (Annapolis: Naval Institute Press, 1998), p. 85, and Adams, *History of the United States*, p. 1013.
25. Henry Adams, ed., *The Writings of Albert Gallatin* (New York: Antiquarian Press, 1960), vol. 1, p. 627 (Gallatin to Monroe, June 13, 1814).
26. Adams, *History of the United States*, p. 1013.
27. Allen C. Clark, *Life and Letters of Dolly Madison* (Washington, D.C.: W. F. Roberts, 1914), p. 166.
28. Ibid.
29. "Ross," *Dictionary of National Biography*, vol. 17, p. 276; Pitch, *Burning of Washington*, pp. 117–20.
30. Pitch, *Burning of Washington*, p. 124.
31. Adams, *History of the United States*, p. 1015.
32. James Ewell, *The Medical Companion*, 3rd ed. (Philadelphia: Anderson & Meecham, 1816), p. 655.
33. Pitch, *Burning of Washington*, p. 142.
34. Ibid., p. 144.

35. Clark, *Life and Letters*, pp. 166–67; Hickey, *War of 1812*, p. 201.

36. Tucker, *Poltroons and Patriots*, p. 584.

37. Arsene Lacarriere Latour, *Historical Memoir of the War in West Florida and Louisiana in 1814–15* (Gainesville, Fla: Historic New Orleans Collection and University Press of Florida, 1999), p. 178 (Monroe to Cochrane, September 6, 1814).

38. Walter Lord, *The Dawn's Early Light* (New York: W. W. Norton, 1972), p. 216.

39. Adams, *History of the United States*, p. 993.

40. Hickey, *War of 1812*, p. 202.

41. Mahan, *Sea Power*, vol. 2, p. 342.

42. Hickey, *War of 1812*, p. 202.

43. Tucker, *Poltroons and Patriots*, p. 742.

44. Antoine Jomini, *The Art of War* (Westport, Conn.: Greenwood Press, n.d.,), p. 349.

O Say, Can You See?

1. Hickey, *War of 1812*, p. 202.

2. Lord, *Dawn's Early Light*, pp. 222–23.

3. Ibid., pp. 224–25.

4. Mahon, *War of 1812*, pp. 307–08; Heidler and Heidler, *Encyclopedia of the War of 1812*, pp. 476–77.

5. Tucker, *Poltroons and Patriots*, pp. 585–89.

6. Gleig, *Campaigns of the British Army*, p. 102.

7. Lord, *Dawn's Early Light*, p. 262.

8. Gleig, *Campaigns of the British Army*, pp. 94–95; Lord, *Dawn's Early Light*, pp. 262–63; Pitch, *Burning of Washington*, pp. 198–200; *Dictionary of National Biography* (hereafter cited as *DNB*), 17, p. 277.

9. *DNB*, 17, p. 277.

10. Gleig, *Campaigns of the British Army*, p. 109.

11. Pitch, *Burning of Washington*, p. 200.

12. Mahon, *War of 1812*, p. 309.

13. Pitch, *Burning of Washington*, p. 188; Heidler and Heidler, *Encyclopedia of the War of 1812*, pp. 12–13.

14. Lord, *Dawn's Early Light*, pp. 274–75; Heidler and Heidler, *Encyclopedia of the War of 1812*, pp. 486–87.

15. Francis Scott, Key, "The Star-Spangled Banner," in *The One Hundred and One Best Songs* (Chicago: Cable Company, 1919), p. 2.

16. Adams, *History of the United States*, p. 1032 (Brooke to Cochrane, September 14, 1814).

17. Mahon, *War of 1812*, p. 311.

18. Heidler and Heidler, *Encyclopedia of the War of 1812*, p. 488.

19. Key, "Star-Spangled Banner," stanza two.

20. Ibid., stanza four.

Still Mr. Madison's War

1. Hickey, *War of 1812*, pp. 237–238.

2. John C. Fredriksen, *War of 1812 Eyewitness Accounts: An Annotated Bibliography* (Westport, Conn.: Greenwood Press, 1997), p. 21.

3. *Annals of the Congress of the United States*, Thirteenth Congress, Third Session, p. 1137 (February 8, 1815).

4. Hickey, *War of 1812*, p. 231.

5. Albert Ellery Bergh, ed. *The Writings of Thomas Jefferson* vol. 14, (Washington: Thomas Jefferson Memorial Association, 1903), p. 216 (Jefferson to Short, November 28, 1814).

6. *United States Gazette*, October 11, 1813.

7. Hickey, *War of 1812*, p. 222.

8. Ibid., pp. 246–47.

9. Adams, *History of the United States*, p. 1091.

10. *Annals of the Congress of the United States*, Thirteenth Congress, Third Session, p. 1105 (January 26, 1815); Hunt, *Writings of James Madison*, p. 319 (Madison to Nicholas, November 25, 1814).

11. Salem (Mass.) *Gazette*, September 23, 1814.

12. Hickey, *War of 1812*, p. 273.

13. *Warren-Adams Letters, Being chiefly a correspondence among John Adams, Samuel Adams, and James Warren*, vol. 2 1778–1814 (Boston: Massachusetts Historical Society, 1925), p. 396.

14. Henry Cabot Lodge, *Life and Letters of George Cabot* (Boston: Little, Brown, 1878), pp. 519, 602.

15. Tucker, *Poltroons and Patriots*, p. 664.

16. Charles M. Wiltse, ed. *The Papers of Daniel Webster: Speeches and Formal Writings*, vol. 1, *1800–1833* (Hanover, N.H.: University Press of New England, 1986), p. 30.

17. Theodore Dwight, *History of the Hartford Convention with a Review of the Policy of the United States Government Which Led to the War of 1812* (New York: N. & J. White, 1833), pp. 377–78.

18. *New York Evening Post*, January 7, 1815.

19. Adams, *History of the United States*, p. 1116.

20. Lodge, *Life and Letters of George Cabot*, p. 602.

21. *Niles' Weekly Register*, 8, no. 6 (April 8, 1815), p. 100.

22. Adams, *History of the United States*, p. 1117.

23. Lodge, *Life and Letters of George Cabot*, pp. 562–63.

24. Harold D. Moser,. et al., eds. *The Papers of Andrew Jackson* vol. 4, 1816–1820 (Knoxville: University of Tennessee Press, 1994), p. 81.

25. Noah Webster, "Letter to Daniel Webster, 1834," *American Historical Review* 9 (October 1903), p. 104.

26. Tucker, *Poltroons and Patriots*, p. 651.

Christmas in Ghent

1. Hickey, *War of 1812*, p. 232.

2. Mahan, *Sea Power*, vol. 2, p. 409.

3. *Annals of the Congress of the United States*, Twelfth Congress, First Session, p. 293 (June 16, 1812).

4. Mahan, *Sea Power*, vol. 2, p. 410.

5. Ibid., pp. 411–13.

6. "Letters of John Quincy Adams," *Massachusetts Historical Society Proceedings* 10 (December 1895), p. 384.

7. Robert V. Remini. *Henry Clay: Statesman for the Union*. (New York: W. W. Norton, 1991), p. 100.

8. "John Quincy Adams to Charles B. Cochran, from Ghent, July 18, 1814," *American Historical Review* 15 (April 1910), p. 574.

9. Bradford Perkins, *Castlereagh and Adams: England and the United States, 1812–1823* (Berkeley: University of California Press, 1964), pp. 58–60.

10. Fred L. Engleman, *The Peace of Christmas Eve* (New York: Harcourt, Brace & World, 1960), p. 96.

11. Mahan, *Sea Power*, vol. 2, p. 414; Remini, *Henry Clay*, pp. 104–05.

12. Mahan, *Sea Power*, vol. 2, p. 414.

13. "John Quincy Adams to Charles B. Cochran," p. 573.

14. Adams, *Writings of Albert Gallatin*, vol. 1, 629.

15. Mahan, *Sea Power*, vol. 2, p. 266.

16. Hopkins, *Papers of Henry Clay*, p. 989.

17. Quoted in Hickey, *War of 1812*, p. 291; Mahan, *Sea Power*, vol. 2, p. 418; Lavender, *Fist in the Wilderness*, p. 220.

18. Hickey, *War of 1812*, p. 284.

19. Ibid., p. 289.

20. Engleman, *Peace of Christmas Eve*, p. 197; Mahon, *War of 1812*, p. 326.

21. Pratt, *History of United States Foreign Policy*, pp. 136–37.

22. Engleman, *Peace of Christmas Eve*, pp. 303–11.

23. Ibid., p. 286; Hickey, *War of 1812*, p. 287.

24. Charles F. Adams, ed., *Memoirs of John Quincy Adams*, vol. 3 (Philadelphia: J. B. Lippincott, 1875). p. 127.

25. Adams, *History of the United States*, p. 1122.

Along the Mighty Mississip'

1. "The Battle of New Orleans," words by Jimmy Driftwood, copyright © Warden Music Co., Inc. Used by permission. All rights reserved.

2. Frank Lawrence Owsley, Jr., *Struggle for the Gulf Borderlands: The Creek War and the Battle of New Orleans, 1812–1815* (Gainesville: University Presses of Florida, 1981), pp. 116–18, 122–26; Mahon, *War of 1812*, pp. 339, 347, 350–51.

3. Owsley, *Struggle for the Gulf Borderlands*, pp. 126–27.

4. Robin Reilly, *The British at the Gates: The New Orleans Campaign in the War of 1812.* (New York: G. P. Putnam's Sons, 1974), pp. 183, 187.

5. Ibid., pp. 192–93; Mahon, *War of 1812*, p. 348.

6. Robert V. Remini, *The Battle of New Orleans: Andrew Jackson and America's First Military Victory* (New York: Penguin Books, 1999), p. 38.

7. Owsley, *Struggle for the Gulf Borderlands*, p. 137.

8. Remini, *Battle of New Orleans*, p. 70.

9. Owsley, *Struggle for the Gulf Borderlands*, pp. 141–45.

10. Remini, *Battle of New Orleans*, p. 89.

11. Ibid., p. 89.

12. Latour, *Historical Memoir of the War*, p. 228 (Jackson to Secretary of War, December 26, 1814).

13. Remini, *Battle of New Orleans*, pp. 93, 125.

14. Ibid., pp. 102, 127–29.

15. Latour, *Historical Memoir of the War*, p. 318 (Thornton to Pakenham, January 8, 1815).

16. Parton, *Life of Andrew Jackson*, vol. 2, pp. 190–91.

17. Remini, *Battle of New Orleans*, pp. 136–38.

18. Parton, *Life of Andrew Jackson*, vol. 2, p. 188.

19. Remini, *Battle of New Orleans*, p. 138.

20. Owsley, *Struggle for the Gulf Borderlands*, p. 131.

21. Remini, *Battle of New Orleans*, p. 132.

22. Ibid., pp. 138–39; Pakenham quote is at Parton, *Life of Andrew Jackson*, vol. 2, p. 192.

23. Parton, *Life of Andrew Jackson*, vol. 2, pp. 190–92.

24. Remini, *Battle of New Orleans*, pp. 140–42.

25. Ibid., p. 144.

26. Ibid., p. 145.

27. Ibid., p. 147.

28. Ibid., p. 148.

29. Ibid., pp. 147–49.

30. Latour, *Historical Memoir of the War*, p. 314 (Lambert to Bathhurst, January 10, 1815).

31. Remini, *Battle of New Orleans*, pp. 150–56.

32. *Ibid.*, pp. 158–61.

33. Latour, *Historical Memoir of the War*, pp. 243–46. (Patterson to Secretary of the Navy, January 13, 1815).

34. Ibid., p. 238 (Jackson to Secretary of War, January 9, 1815).

35. Ibid., p. 318 (Thornton to Pakenham, January 8, 1815; Return of Ordnance, January 8, 1815).

36. Ibid., p. 314 (Lambert to Bathhurst, January 10, 1815); Remini, *Battle of New Orleans*, p. 164.

37. Latour, *Historical Memoir of the War*, p. 239 (Jackson to Secretary of War, January 13, 1815).

38. Ibid., p. 241 (Jackson to Secretary of War, January 19, 1815).

39. Harold D. Moser, et al., eds., *The Papers of Andrew Jackson*, vol. 3, 1814–1815, (Knoxville: University of Tennessee Press, 1991), p. 252 (Jackson to Winchester, January 19, 1815).

40. Owsley, *Struggle for the Gulf Borderlands*, pp. 170–74; Latour, *Historical Memoir of the War*, p. 261 (Cochrane to Jackson, February 13, 1815).

41. "The Battle of New Orleans," words by Jimmy Driftwood, copyright © Warden Music Co., Inc. Used by permission. All rights reserved.

42. Owsley, *Struggle for the Gulf Borderlands*, p. 168.

A Nation at Last

1. Engleman, *Peace of Christmas Eve*, pp. 287–88 (*London Times*, December 30, 1814).

2. Ibid., p. 288.

3. Ibid., pp. 289–90.

4. Hickey, *War of 1812*, p. 279.

5. Engelman, *Peace of Christmas Eve*, pp. 287, 290.

6. *Annals of the Congress of the United States*, Thirteenth Congress, Third Session, p. 255 (February 18, 1815).

7. *Niles' Weekly Register* 8, no. 25 (February 18, 1815), p. 1.

8. Reilly, *British at the Gates*, p. 335; Owsley, *Struggle for the Gulf Borderlands*, p. 178.

9. Robert L. Meriwether, ed., *The Papers of John C. Calhoun*, vol. 1, 1801–1817 (Columbia: University of South Carolina Press, 1959), p. 96 (Calhoun to Noble, March 22, 1812).

10. Eisenhower, *Agent of Destiny*, p. 103.

11. Remini, *Battle of New Orleans*, p. 198.

12. Reilly, *British at the Gates*, p. 329.

13. Margaret L. Callcott, ed., *Mistress of Riversdale: The Plantation Letters of Rosalie Stier Calvert, 1795–1821* (Baltimore: Johns Hopkins University Press, 1991), p. 278.

14. *Annals of the Congress of the United States*, Thirteenth Congress, Second Session, p. 544 (December 7, 1813).

15. Hamilton, *Writings of James Monroe*, p. 346 (Monroe to Jackson, December 16, 1816).

16. Dexter Perkins, *A History of the Monroe Doctrine* (Boston: Little, Brown, 1963), p. 29.

17. George Dangerfield, *The Awakening of American Nationalism, 1815–1828* (New York: Harper & Row, 1965), p. 2.

18. Latour, *Historical Memoir of the War*, p. 337 (Jackson to Command, January 21, 1815).

19. Remini, *Andrew Jackson and the Course*, p. 166.

20. Hunt, *Writings of James Madison*, p. 193 (Madison to Congress, June 1, 1812); *Annals of the Congress of the United States*, Twelfth Congress, First Session, p. 2322 (June 18, 1812).

21. Adams, *Writings of Albert Gallatin*, pp. 651–52 (Gallatin to Jefferson, September 6, 1815).

22. Ibid., p. 700 (Gallatin to Lyon, May 7, 1816).

BIBLIOGRAPHY

Books

Adams, Henry. *History of the United States of America During the Administrations of James Madison.* New York: The Library of America, 1986. Originally published as part of *History of the United States of American during the Administrations of Jefferson and Madison,* 1889–1891.

Ball, Charles. *A Narrative of the Life and Adventures of Charles Ball, A Black Man.* New York: John S. Taylor, 1837.

Berton, Pierre. *Flames Across the Border, 1813–1814.* Toronto: McClelland and Stewart, 1981.

———.*The Invasion of Canada, 1812–1813.* Toronto: McClelland and Stewart, 1980.

Burstein, Andrew. *The Passions of Andrew Jackson.* New York: Alfred A. Knopf, 2003.

Caffrey, Kate. *The Twilight's Last Gleaming: Britain vs. America, 1812–1815.* New York: Stein and Day, 1977.

Callcott, Margaret L., ed. *Mistress of Riversdale: The Plantation Letters of Rosalie Stier Calvert, 1795–1821.* Baltimore: Johns Hopkins University Press, 1991.

Clark, Allen C. *Life and Letters of Dolly Madison.* Washington, D.C.: W. F. Roberts, 1914.

Cleaves, Freeman. *Old Tippecanoe: William Henry Harrison and His Time.* New York: Charles Scribner's Sons, 1939.

Coles, Harry L. *The War of 1812*. Chicago: University of Chicago Press, 1965.

Cruikshank, E. *The Documentary History of the Campaign upon the Niagara Frontier in the Year 1813*. Welland, Ontario: Lundy's Lane Historical Society, 1902.

Dangerfield, George. *The Awakening of American Nationalism, 1815–1828*. New York: Harper & Row, 1965.

———. *The Era of Good Feelings*. New York: Harcourt, Brace and Company, 1952.

Dillon, Richard. *We Have Met the Enemy: Oliver Hazard Perry: Wilderness Commodore*. New York: McGraw Hill, 1978.

Dwight, Theodore. *History of the Hartford Convention with a Review of the Policy of the United States Government Which Led to the War of 1812*. New York: N. & J. White, 1833.

Edmunds, R. David. *Tecumseh and the Quest for Indian Leadership*. New York: Little, Brown, 1984.

Eisenhower, John S. D. *Agent of Destiny: The Life and Times of General Winfield Scott*. New York: Free Press, 1997.

Engleman, Fred L. *The Peace of Christmas Eve*. New York: Harcourt, Brace & World, 1960.

Ewell, James. *The Medical Companion*. 3rd ed. Philadelphia: Anderson & Meecham, 1816.

Fitch, Raymond E., ed. *Breaking with Burr: Harman Blennerhassett's Journal, 1807*. Athens: Ohio University Press, 1988.

Fitz-Enz, David G. *The Final Invasion: Plattsburgh, the War of 1812's Most Decisive Battle*. New York: Cooper Square Press, 2001.

Fleming, Thomas. *Duel: Alexander Hamilton, Aaron Burr and the Future of America*. New York: Basic Books, 1999.

Forester, C. S. *The Age of Fighting Sail: The Story of the Naval War of 1812*. Garden City, N.Y.: Doubleday, 1956.

Fredriksen, John C. *War of 1812 Eyewitness Accounts: An Annotated Bibliography*. Westport, Conn.: Greenwood Press, 1997.

Garraty, John A. *The American Nation: A History of the United States*. New York: Harper & Row, 1966.

Gleig, G. R. *The Campaigns of the British Army at Washington and New Orleans*. Totowa, N.J.: Rowman and Littlefield, 1972. Originally published by John Murray, London, 1827.

Grant, Bruce. *Isaac Hull: Captain of Old Ironsides*. Chicago: Pellegrini and Cudahy, 1947.

Graves, Donald E. *Field of Glory: The Battle of Crysler's Farm, 1813.* Toronto: Robin Brass Studio, 1999.

———. *Where Right and Glory Lead! The Battle of Lundy's Lane, 1814.* Toronto: Robin Brass Studio, 1997.

Hansen, Harry. *Old Ironsides: The Fighting Constitution.* New York: Random House, 1955.

Heidler, Jeanne T., and David S. Heidler, eds. *Encyclopedia of the War of 1812.* Santa Barbara, Calif.: ABC-Clio, 1997.

Hendrickson, Robert. *Hamilton*, vol. 2 (1789–1804). New York: Mason/Charter, 1976.

Hickey, Donald R. *The War of 1812: A Forgotten Conflict.* Urbana: University of Illinois Press, 1989.

Hill, Charles E. "James Madison," in Samuel Flagg Bemis, ed. *The American Secretaries of State and Their Diplomacy*, vol. 3, New York: Cooper Square Publishers, Inc., 1963. Originally published by Alfred A. Knopf, 1928.

Hitsman, J. Mackay. *The Incredible War of 1812: A Military History.* Toronto: Robin Brass Studio, 1999. This edition updated by Donald E. Graves. Originally published by University of Toronto Press, 1965.

Hull, William. *Memoirs of the Campaign of the North Western Army of the United States, A.D. 1812.* Boston: True & Greene, 1824.

Jacobs, James R. *Tarnished Warrior: The Story of Major-General James Wilkinson.* New York: Macmillan, 1938.

Jomini, Antoine. *The Art of War.* Westport, Conn.: Greenwood Press, n.d. Originally published by J. B. Lippincott, 1862.

Latour, Arsene Lacarrière. *Historical Memoir of the War in West Florida and Louisiana in 1814–15.* Gainesville, Fla.: Historic New Orleans Collection and University Press of Florida, 1999, with an introduction by and edited by Gene A. Smith. Originally published by John Conrad and Company, Philadelphia 1816.

Lavender, David. *The Fist in the Wilderness.* Garden City, N.Y.: Doubleday, 1964.

Lodge, Henry Cabot. *Life and Letters of George Cabot.* Boston: Little Brown, 1878.

Lomask, Milton. *Aaron Burr: The Conspiracy and Years of Exile, 1805–1836.* New York: Farrar, Straus, Giroux, 1982.

Lord, Walter. *The Dawn's Early Light.* New York: W. W. Norton, 1972.

Lossing, Benson J. *The Pictorial Field-Book of the War of 1812.* New York: Harper & Brothers, 1869.

Lowther, Minnie Kendall. *Blennerhassett Island in Romance and Tragedy*. Rutland, Vt.: Tuttle Publishing, 1939.

McAfee, Robert Breckinridge. *History of the Late War in the Western Country*, Ann Arbor, Michigan: Unversity Microfilms, 1966. Originally published by Worsley & Smith, Lexington, Kentucky, 1816.

Macdonough, Rodney. *The Life of Commodore Thomas Macdonough, U.S. Navy*. Boston: Fort Hill Press, 1909.

Mackenzie, Alexander Slidell. *The Life of Commodore Oliver Hazard Perry*. New York: Harper & Brothers, 1840.

Maclay, Edgar Stanton. *A History of the United States Navy*, vol. 1. New York: D. Appleton, 1910.

Mahan, A. T. *Sea Power in Its Relations to the War of 1812*. Boston: Little, Brown, 1905.

Mahon, John K. *The War of 1812*. Gainesville: University of Florida Press, 1972.

Malcomson, Robert. *Lords of the Lake: The Naval War on Lake Ontario, 1812–1814*. Annapolis: Naval Institute Press, 1998.

Malone, Dumas. *Jefferson and His Time: The Sage of Monticello*. Boston: Little, Brown, 1977.

———. *Jefferson the President: Second Term, 1805–1809*. Boston: Little, Brown, 1974.

Muller, Charles G. *The Proudest Day: Macdonough on Lake Champlain*. New York: John Day, 1960.

Nettles, Curtis P. *The Emergence of a National Economy, 1775–1815*, vol. 3 of *The Economic History of the United States*. New York: Holt, Rinehart and Winston, 1962.

Owsley, Frank Lawrence Jr. *Struggle for the Gulf Borderlands: The Creek War and the Battle of New Orleans, 1812–1815*. Gainesville: University Presses of Florida, 1981.

Parton, James. *Life of Andrew Jackson*. New York: Mason Brothers, 1860.

Perkins, Bradford. *Castlereagh and Adams: England and the United States, 1812–1823*. Berkeley: University of California Press, 1964.

———. *Prologue to War: England and the United States, 1805–1812*. Berkeley: University of California Press, 1963.

Perkins, Dexter. *A History of the Monroe Doctrine*. Boston: Little, Brown, 1963.

Pitch, Anthony S. *The Burning of Washington: The British Invasion of 1814*. Annapolis: Naval Institute Press, 1998.

Pratt, Julius W. *A History of United States Foreign Policy*. Englewood Cliffs, N.J.: Prentice-Hall, 1955.

Ramsay, Jack C., Jr. *Jean Laffite: Prince of Pirates*. Austin, Texas: Eakin Press, 1996.

Reilly, Robin. *The British at the Gates: The New Orleans Campaign in the War of 1812*. New York: G. P. Putnam's Sons, 1974.

Remini, Robert V. *Andrew Jackson and His Indian Wars*. New York: Viking, 2001.

———. *Andrew Jackson and the Course of American Empire, 1767–1821*. New York: Harper & Row, 1977.

———. *The Battle of New Orleans: Andrew Jackson and America's First Military Victory*. New York: Penguin Books, 1999.

———. *Henry Clay: Statesman for the Union*. New York: W. W. Norton, 1991.

Roosevelt, Theodore. *The Naval War of 1812*. Annapolis: Naval Institute Press, 1987. Originally published by G. P. Putnam's Sons, 1882.

Scott, Winfield. *Memoirs of Lieut.-General Scott, LL.D.* Freeport, N.Y.: Books for Libraries Press, 1970. Originally published in 1864.

Skaggs, David Curtis, and Gerard T. Altoff. *A Signal Victory: The Lake Erie Campaign, 1812–1813*. Annapolis: Naval Institute Press, 1997.

Sugden, John. *Tecumseh: A Life*. New York: Henry Holt, 1998.

Tucker, Glenn. *Poltroons and Patriots: A Popular Account of the War of 1812*. Indianapolis: Bobbs-Merrill, 1954.

Watson, Paul Barron. *The Tragic Career of Commodore James Barron*. New York: Coward-McCann, 1942.

Whitaker, Arthur Preston. *The Spanish-American Frontier, 1783–1795*. Lincoln: University of Nebraska, 1969. Reprint of 1927 edition.

Wilkinson, James. *Memoirs of My Own Times*. Philadelphia: Abraham Small, 1816.

Articles

Austin, Catherine, ed. "Letters of Elbridge Gerry, Jr." *Massachusetts Historical Society Proceedings* 47 (June 1914), pp. 480–528.

"John Quincy Adams to Charles B. Cochran, from Ghent, July 18, 1814." *American Historical Review* 15 (April 1910), p. 574.

"Letters of John Quincy Adams," *Massachusetts Historical Society Proceedings* 10 (December, 1895) pp. 374–92.

R. B. McAfee, "William H. Harrison," *The Library of Historic Characters and Famous Events* (Boston J. B. Millet, 1907), vol. 11, pp. 309–36.

Taylor, John M. "An Accomplished Villain." *American History Illustrated* 13, no. 9 (January 1979) pp. 4–9, 47–49.

Webster, Noah. "Letter to Daniel Webster, 1834." *American Historical Review* 9 (October 1903), p. 104.

Papers and Government Documents

Adams, Charles F., ed. *Memoirs of John Quincy Adams*, vol. 3. Philadelphia: J. B. Lippincott, 1874.

Adams, Henry, ed. *The Writings of Albert Gallatin*, vol. 1. New York: Antiquairan Press, 1960. Originally published in 1879 by: J. B. Lippincott, Philadelphia.

Annals of the Congress of the United States. Eleventh, Twelfth, and Thirteenth Congresses.

Bergh, Albert Ellery, ed. *The Writings of Thomas Jefferson*, vol. 14. Washington D.C.: Thomas Jefferson Memorial Association, 1903.

Brunson, Alfred. *A Western Pioneer, or, Incidents in the Life and Times of Alfred Brunson*. 2 vols. Cincinnati: Hitchcock, Walden, 1872.

Bureau of American Ethnology. *Annual Report*, 1897–1898. Washington D.C.: Government Printing Office, 1900.

Dudley, William S., ed. *The Naval War of 1812: A Documentary History*, vol. 1, 1812. Washington, D.C.: Naval Historical Center, 1985.

———. *The Naval War of 1812: A Documentary History*, vol. 2, 1813. Washington, D.C.: Naval Historical Center, 1992.

Esarey, Logan, ed. *Messages and Letters of William Henry Harrison*, vol. 1, 1800–1811. Indianapolis: Indiana Historical Commission, 1923.

Hamilton, Stanislaus Murray, ed. *The Writings of James Monroe*, vol. 5, 1807–1816. New York: G. P. Putnam's Sons, 1901.

Hastings, Hugh, ed. *Public Papers of Daniel D. Tompkins, Governor of New York, 1807–1817*, Military–vol. 3. Albany: J. B. Lyon, 1902.

Hopkins, James F., ed. *The Papers of Henry Clay*, vol. 1., The Rising Statesman, 1797–1814. Lexington: University of Kentucky Press, 1959.

Hunt, Gaillard, ed. *The Writings of James Madison*, vol. 8, 1808–1819. New York: G. P. Putnam's Sons, 1908.

Kline, Mary-Jo, ed. *Political Correspondence and Public Papers of Aaron Burr*, vol. 2. Princeton, N.J.: Princeton University Press, 1983.

Meriwether, Robert L., ed. *The Papers of John C. Calhoun*, vol. 1, 1801–1817. Columbia: University of South Carolina Press, 1959.

Moser, Harold D., et al., eds. *The Papers of Andrew Jackson*, vol. 3, 1814–1815. Knoxville: University of Tennessee Press, 1991.

———. *The Papers of Andrew Jackson*, vol. 4, 1816–1820. Knoxville: University of Tennessee Press, 1994.

Stagg, J. C. A. *The Papers of James Madison*. Presidential Series, vol. 4, November 5, 1811–July 9, 1812. Charlottesville: University Press of Virginia, 1999.

Syrett, Harold C., ed. *The Papers of Alexander Hamilton*, vol. 23, April 1799–October 1799. New York: Columbia University Press, 1976.

———. *The Papers of Alexander Hamilton*, vol. 25, July 1800–April 1802. New York: Columbia University Press, 1977.

———. *The Papers of Alexander Hamilton*, vol. 26, May 1, 1802–October 23, 1804. New York: Columbia University Press, 1979.

Warren-Adams Letters, Being Chiefly a Correspondence among John Adams, Samuel Adams, and James Warren, vol. 2, 1778–1814. Boston: Massachusetts Historical Society, 1925.

Wiltse, Charles M., ed. *The Papers of Daniel Webster: Speeches and Formal Writings*, vol. 1, 1800–1833. Hanover, N.H.: University Press of New England, 1986.

Newspapers

Caledonian Mercury (Edinburgh).

Columbian Centinel (Boston).

Connecticut Courant(Hartford).

Federal Republican (Georgetown).

London Morning Chronicle.

London Times.

Niles' Weekly Register.

New York Gazette.

New York Post.

New York Evening Post.

Salem Gazette (Massachusetts).

United States Gazette.

INDEX

Page numbers in *italics* refer to maps.

P.S.

Insights,
Interviews
& More . . .

Meet
Walter R. Borneman

© 2004 Marlene M. Whyte

WALTER R. BORNEMAN has written books
and articles about mountains, railroads, and
the American West—most recently, *Alaska:
Saga of a Bold Land* (HarperCollins, 2003)
and *1812: The War That Forged a Nation*
(HarperCollins, 2004). In his home state of
Colorado, he is best-known as the coauthor
of *A Climbing Guide to Colorado's Fourteeners*,
first published in 1978 and in print for
twenty-five years. Borneman has worked for
the Colorado Historical Society, practiced
law with frequent involvement in historic
preservation issues, and served as the first
chairman of the Colorado Fourteeners
Initiative—a nonprofit organization devoted
to the stewardship and preservation of the
state's highest peaks. He is the president of the
Walter V. and Idun Y. Berry Foundation, which
funds postdoctoral fellowships in children's
health at Stanford University.

For more information about Walter R.
Borneman and his books, please visit
www.walterborneman.com.

Confessions of a
Childhood Historian

I NEVER FULLY UNDERSTOOD why none of my children liked history. Their moans of "boring" and "who needs to know that" just doesn't square with my childhood experiences. Back then, the fall of 1960 found me enrolled in Mrs. Burkholder's third-grade class in Richfield, Ohio. We were a mature bunch of eight-year-olds who ran around the playground shouting slogans that we didn't quite understand from the heated Kennedy-Nixon campaign then in full swing. But what really held my fascination that year—thanks in no small measure to Mrs. Burkholder's encouragement— was the impending centennial of the Civil War. Now there was something that an eight-year-old could really sink his teeth into.

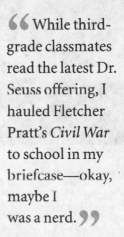

First desk, Christmas 1957

Call me a nerd if you will, but I simply couldn't get enough of the images of Fort Sumter, Pickett's Charge, or Lee and Grant meeting on the steps at Appomattox. While third-grade classmates read the latest Dr. Seuss offering, I hauled Fletcher Pratt's *Civil War* to school in my briefcase—okay, maybe I was a nerd. I built cardboard models of the *Monitor* and the *Merrimac* for a Cub Scout project and "published" my first book, an eight-page Crayola-colored extravaganza of the conflict. But the Civil War was just the beginning. ▶

> ❝ While third-grade classmates read the latest Dr. Seuss offering, I hauled Fletcher Pratt's *Civil War* to school in my briefcase—okay, maybe I was a nerd. ❞

Confessions of a Childhood Historian
(continued)

My brother and I prowled through the oak and maple forests surrounding our house in the roles of our favorite Walt Disney heroes, Francis Marion, that daring "swamp fox" of the American Revolution, and Peter Horry, his trusted lieutenant. *The Last of the Mohicans* was my most-worn comic book long before I had an inkling what the French and Indian War was all about.

Author, little brother Bruce, and "Gram," Gettysburg 1959

When it was time to cut and paste flags in art class, I was not content to craft Ohio's pennant-shaped ensign from colored paper. For me, it had to be Oliver Hazard Perry's victory banner. Thus, my blue flag with its white cut-out letters of "Don't Give Up the Ship" joined twenty-some others in a row above the blackboard. Later, Mrs. B. confided to me that on more than one occasion amid the commotion of third-graders, she had looked up to those words and found a renewed resolve.

Maybe one reason that history was never boring for me as a child was that in my imagination it always gave us something to do. When we weren't worrying about the Cleveland Indians winning the pennant—*that* didn't take long in those days—history gave us a hundred roles to play.

As I have written about America's past, I

have tried to remember the excitement I felt for those topics long ago. Sound scholarship need not be dull. Indeed, as I am fond of saying, the true events and characters are frequently much more compelling than even the most spell-binding fiction. So call me a third-grade nerd if you will, but gosh, it was fun being one! ❧

> **❝ Sound scholarship need not be dull. ❞**

What about **Canada?**

TWO WEEKS before *1812* was first published, I gave the keynote address to the Alaska Historical Society's annual meeting and settled down to dinner expecting some stimulating conversation about Alaska. The conversation was lively all right, but what everyone wanted to talk about was not Alaska, but rather the War of 1812. It helped, of course, to have a Canadian at the table who adamantly declared that the Canadians—with some assistance from their British landlords—had won the war. Militarily, at least, I told her that she definitely had a point. After all, Henry Clay's boasts and the attacks of at least eight American armies over three years had failed to make so much as one permanent dent in Canada's borders.

All of which begs the question, if the War of 1812 was indeed the war that forged the American nation, what role did it play in Canada's evolution into an independent country?

Pierre Berton, arguably Canada's greatest historian, had no doubts about what the War of 1812 meant to Canada, or for that matter, how to characterize it. Berton called the first volume of his history of the war *The Invasion of Canada: 1812–1813,* and argued that without the war, at least Ontario and points west would likely have become American. Echoing this theme, a Canadian journalist reviewing *1812* in 2004 noted that while I was clearly referring to the United States as the country forged by the war, I might "just as easily be referring to Canada." Finally, I have been amazed at the passionate views flying about contemporary War of 1812 websites over the topic of "who won the war."

> ❝ If the War of 1812 was indeed the war that forged the American nation, what role did it play in Canada's evolution into an independent country? ❞

Hopefully, those finishing my history will recognize that my assertion of the "forging of a nation" results far less from scattered military victories—save perhaps Jackson's at New Orleans—and much more from the collective boasts of a nation that somehow managed to persevere without those victories. But what about Canada? Did the War of 1812 forge the Canadian nation in the same sense? Most Americans in 1812 were genuinely surprised by the stout defense that Canadians put up—both with and without British troops. In doing so, Canadians inexorably allied themselves with Great Britain and turned their backs on political union with the United States. Through their resistance, Canadians embarked on a completely separate journey toward independence that would take more than a century and generally lack the violence with which their American cousins had departed the British Empire.

In the wake of the War of 1812, Canada went through a number of governmental reorganizations—some of which were related to the historic fact that a good half of the country had its roots in Franco rather than Anglo traditions. Two short-lived rebellions in 1837 and 1838 convinced Queen Victoria to dispatch Lord Durham to Canada to report on colonial grievances and suggest ways to resolve them. Durham deduced—amid no small amount of British criticism—that the only way to keep Canada in the British Empire was to grant it some measure of self-government.

The British North American Act of 1867 finally established the Dominion of Canada ▶

> **"** Through their resistance, Canadians embarked on a completely separate journey toward independence that would take more than a century and generally lack the violence with which their American cousins had departed the British Empire. **"**

What about Canada? *(continued)*

with the provinces of Quebec, Ontario, Nova Scotia, and New Brunswick. From this foundation, additional provinces were added across North America above the forty-ninth parallel. (British Columbia became the sixth province in 1871 after a gold rush fueled its growth in a manner similar to California's and the national government declared its intention to build a Canadian transcontinental railroad.) But it was not until the British Parliament passed the Statute of Westminster in 1931 that the Dominion of Canada was truly and completely independent, but a voluntary partner in the British Commonwealth.

Did the War of 1812 forge the Canadian nation in the same sense that I claim it forged the American nation? I don't think so. To be sure, the war proved that Canada was not a plum ripe for dropping into the American union. Canada's resistance to American persuasions, as well as its arms, also doubtless preserved the opportunity to forge an independent nation, but it would take some time to accomplish this.

What is so wonderful about history is that we can continue to debate events of long ago and their ramifications with the same vigor as the original participants. To paraphrase one Canadian's recent views on a War of 1812 website: "The British 'think' they won the war because the Americans sued for peace. . . . The American 'think' they won the war because of the battle of New Orleans. . . . We Canadians 'know' we won the war because we are still here almost two hundred years later." Hurrah for the maple leaf! ~

66 What is so wonderful about history is that we can continue to debate events of long ago and their ramifications with the same vigor as the original participants. 99

The "Extras" File

IN PRECOMPUTER DAYS my writing projects always had an unruly file of "extras." These were scraps of paper laden with salient facts, pithy quotes, and insightful commentary that I had come across, but did not use. Writing with a computer has made managing the "extras" file much, much easier. Don't want something in there, but too good to throw away because you know if you do, you'll want it again? Easy. A quick cut-and-paste and it's secure in the "extras" file—as long as the hard drive lasts. Here are a few samples from my "extras" file on *1812: The War That Forged a Nation.*

• Isaac Hull's biographer characterized the importance of the *Constitution*'s first victory just after General William Hull had surrendered Detroit thusly: "Defeats on land were forgotten," Bruce Grant wrote, "and the splintered hulk of the *Guerriere* on the bottom of the ocean was worth more in the public mind than the whole province General Hull had so ingloriously tossed into the hands of the British." [Grant, *Hull,* p. 5]

• Did you know that in 1815 when the U.S. Navy built two new 44-gun frigates, they were christened *Guerriere* and *Java* after the vanquished foes of *Old Ironsides*? Naming new warships after defeated enemy vessels was a tradition that the young U.S. Navy quickly thought better of.

• Here's historian Glenn Tucker's take on the Hartford Convention: "The leaders who still schemed for secession in late 1814 were like ▶

> 66 [The unruly file of 'extras'] were scraps of paper laden with salient facts, pithy quotes, and insightful commentary that I had come across, but did not use. 99

The "Extras" File *(continued)*

retired generals, grumbling over strategy and unaware that their soldiers were no longer behind them but were marching with new captains intent on a fresh campaign."
[Tucker, *Poltroons and Patriots,* p. 651]

• Sometimes the veil of history suggests scenes in black-and-white, but this description of the Fifth Maryland Volunteer Infantry at the battle of Blandensburg proves otherwise. "Recruited from the cream of Baltimore society, they were resplendent in blue jackets trimmed in red, white pantaloons, black gaiters, white cross-belts, and heavy leather helmets topped with one black plume and one red one." [Mahon, *War of 1812,* pp. 297–298]

• "Last Guns at Sea"—the great battle at New Orleans was not the only engagement fought after the signing of the peace treaty. Initially, I planned "Last Guns at Sea" to follow the chapter on the Battle of New Orleans, but concluded that this would be anticlimactic and a definite disruption in the flow to the final chapter. Nonetheless, there are two great stories here.

In January 1815, before word of the peace arrived, Commodore Stephen Decatur slipped out of New York harbor in command of the frigate USS *President* and tried to run the British blockade. In a blinding snowstorm that momentarily blew a British squadron of four frigates off station, *President* ran aground off Sandy Hook. Although refloated within two hours, the vessel suffered considerable damage. When a strong west gale kept Decatur from returning to the safety of New York harbor, he steered eastward along Long Island

> 66 Sometimes the veil of history suggests scenes in black-and-white, but this description of the Fifth Maryland Volunteer Infantry at the battle of Blandensburg proves otherwise. 99

and was trapped by the returning British ships. *President* acquitted herself well and in a running duel took HMS *Endymion* out of action. But in the end, cornered by two other British frigates, proud Decatur of the fight with the *Macedonian* was forced to surrender. This defeat might have cost Decatur much of his reputation had news of the peace not reduced the defeat to an afterthought. [Roosevelt, *The Naval War of 1812*, pp. 357–365]

The *Constitution* fared better. Seven days before the Treaty of Ghent was signed, *Constitution* stole out of Boston harbor bound for the eastern Atlantic. On February 20, 1815, some 200 miles off Madeira, *Old Ironsides* met two smaller ships, HMS *Cyane* and HMS *Levant*. Rather than flee, the British captains turned to engage. The combined broadside weight of the British ships slightly exceeded *Constitution*'s, but it was composed mostly of carronades. By the time the ships had closed, the *Constitution* had brought her long guns to bear on first *Cyane* and then *Levant* and the British ships struck their colors. [Roosevelt, *The Naval War of 1812*, pp. 372–378] ∽

. . . And Don't Forget the Upcoming **Bicentennial Celebrations**

PLANS ARE UNDER WAY in the United States and Canada for a variety of celebrations, reenactments, and exhibits that will commemorate the two hundredth anniversary of the War of 1812. The U.S. Army Center of Military History, U.S. National Park Service, and Parks Canada are among the organizations involved.

Beginning in 2012, commemorations will include events at Queenston Heights, Fort George, Horseshoe Bend, Lundy's Lane, and Fort McHenry, culminating with the two hundredth anniversary of the Battle of New Orleans on January 8, 2015. ◡

What's Next?
Borneman Takes On the French and Indian War

Walter R. Borneman's forthcoming historical book on the French and Indian War will be published in hardcover from HarperCollins in Fall 2006.

MY STRONG INCLINATION after writing *1812* was to look westward—both geographically and chronologically. Yet as I pondered my next move, I found myself drawn to events a generation *before* the American Revolution with the same fascination that I had just written about events a generation *after* it. Here was a period that had decided the fate of the entire North American continent—not just between England and France, but among Spain and Native Americans as well.

From the Mississippi and Ohio rivers, to the falls of Niagara, and down the St. Lawrence, North America's frontiers erupted in flames and unleashed escalating rivalries into what quickly became history's truly first *world* war. By the time hostilities concluded, the war had been fought not only in North America but also on European battlefields and in colonies around the globe.

In North America, the war was characterized by epic wilderness treks by Rogers' Rangers, dogged campaigns to capture strategic lynchpins such as Louisbourg and Fort Ticonderoga, full-scale naval engagements in the Caribbean, and the legendary battle of Quebec atop the Plains of Abraham. Frequently overlooked is that just when the English thought they had won ▶

half a continent, France counterattacked and almost recaptured Quebec.

When the warring powers finally met to sign the 1763 Treaty of Paris, the map of the world looked quite different than it had seven years before. As historian Francis Parkman succinctly put it, "half a continent changed hands at the scratch of a pen." But the triumphs of one war sowed the seeds of discontent that would lead to another. England had indeed won a continent, but in doing so, it had also lit the fuse of what would become the American Revolution. ∿

66 As historian Francis Parkman succinctly put it, 'half a continent changed hands at the scratch of a pen.' 99

Have You Read?
More by
Walter R. Borneman

ALASKA: SAGA OF A BOLD LAND

From Russian fur traders to the gold rush, extraordinary railroads, World War II, the oil boom, and the fight over the Arctic National Wildlife Refuge (ANWR), this is the most complete history of Alaska.

"This is narrative history told in superlatives."
—David Lavender

"Just plain terrific."   —Bradford Washburn

The Web Detective

*Check out the following websites for more information about
the War of 1812*

http://www.flaghouse.org
*for information about the Star-Spangled Banner Museum,
Baltimore, Maryland*

http://www.patriotsoffortmchenry.org
for information about the Fort McHenry, Baltimore, Maryland

http://www.pc.gc.ca/lhn-nhs/on/queenston/index_e.asp
*for information about the Queenston Heights National Historic Site
of Canada, Niagara-on-the-Lake, Ontario*

http://www.nps.gov.jela
*for information about the Jean Lafitte National Historical Park
and Preserve, New Orleans, Louisiana*

http://www.usdaughters1812.org
*for information about the National Society of the United States
Daughters of 1812*

http://www.societyofthewarof1812.org
for information about the General Society of the War of 1812

D on't miss the next book by your favorite author. Sign up now for AuthorTracker by
visiting www.AuthorTracker.com.